DESIGNS

DESIGNS

FOR SCIENCE LITERACY

PROJECT 2061

AMERICAN ASSOCIATION FOR THE ADVANCEMENT OF SCIENCE

OXFORD UNIVERSITY PRESS

NEW YORK   OXFORD

2000

OXFORD UNIVERSITY PRESS

Oxford   New York

ATHENS   AUCKLAND   BANGKOK   BOGOTA   BOMBAY
BUENOS AIRES   CALCUTTA   CAPE TOWN   DAR ES SALAAM
DELHI   FLORENCE   HONG KONG   ISTANBUL
KARACHI   KUALA LUMPUR   MADRAS   MADRID
MELBOURNE   MEXICO CITY   NAIROBI   PARIS
SINGAPORE   TAIPEI   TOKYO   TORONTO

AND ASSOCIATED COMPANIES IN

BERLIN   IBADAN

Library of Congress Cataloging-in-Publication Data

Project 2061 (American Association for the Advancement of Science)

Designs for Science Literacy.

p.   cm.

Includes bibliographical references and index.

ISBN 0-19-513278-5

1. Science—Study and teaching—United States.   2. Mathematics—Study and teaching—United States.
3. Engineering—Study and teaching—United States.   I. American Association for the Advancement of Science.

Q183.3.A1D48   2000

507'.1'073—dc21

99-38299

CIP

Designs on Disk contains Macromedia Shockwave™ Player and Macromedia Flash™ Player software by Macromedia, Inc.,
Copyright © 1995-1999, Macromedia, Inc. All rights reserved. Macromedia, Shockwave, and Flash are trademarks of Macromedia, Inc.

The following are referenced in *Designs for Science Literacy* and *Designs on Disk:*
Acrobat—a trademark of Adobe Systems, Inc.   Macintosh—a trademark of Apple Computer, Inc.
Netscape Communicator—a trademark of Netscape communications Corporation
Shockwave and Flash—trademarks of Macromedia, Inc.   Windows—a trademark of Microsoft Corporation

1   3   5   7   9   8   6   4   2
Printed in the United States of America
on acid-free paper

CONTENTS

Founded in 1848, the **American Association for the Advancement of Science** (AAAS) is the world's largest federation of scientific and engineering societies, with nearly 300 affiliate organizations. In addition, AAAS counts more than 143,000 scientists, engineers, science educators, policy makers, and interested citizens among its individual members, making it the largest general scientific organization in the world. The Association's goals are to further the work of scientists, facilitate cooperation among them, foster scientific freedom and responsibility, improve the effectiveness of science in the promotion of human welfare, advance education in science, and increase public understanding and appreciation of the importance and promise of the methods of science in human progress.

Project 2061 is a long-term initiative of AAAS to reform K-12 education in natural and social science, mathematics, and technology. Begun in 1985, Project 2061 is developing a comprehensive set of science education reform tools and provides workshops and other professional development services to help educators make science literacy a reality for all American students.

The AAAS wishes to express its gratitude to the following for their generous support of Project 2061:

❋

CARNEGIE CORPORATION OF NEW YORK
HEWLETT-PACKARD COMPANY
JOHN D. AND CATHERINE T. MACARTHUR FOUNDATION
ANDREW W. MELLON FOUNDATION
ROBERT N. NOYCE FOUNDATION
THE PEW CHARITABLE TRUSTS
NATIONAL SCIENCE FOUNDATION

❋

PREFACE

S*cience for All Americans*, Project 2061's seminal report published in 1989, deals only with learning goals—what students should understand and be able to do after they have left school as a result of their total school experience—but not with how to organize the curriculum to achieve these goals. Now, with the publication of *Designs for Science Literacy*, the curriculum takes center stage.

Why should it have taken so long? One reason was that Project 2061 was crafting an entire set of interrelated tools to enable educators to realize the promise of *Science for All Americans*. But, truth be told, it has also been a struggle to create *Designs*. The struggle was not the usual one of securing funds enough to proceed, but rather a strategic and conceptual one. Was it the job of Project 2061 to create an entire K-12 curriculum that would result in all students achieving the goals set out in *Science for All Americans*? We thought that to be far too ambitious and inconsistent with a healthy diversity of curricula. But if not that, what? Gradually, out of extensive discussions of staff, advisors, and consultants emerged a strong conviction, supported by our advisory body the National Council on Science and Technology Education, that the project should tackle the fundamental challenge of *how* to design entire K-12 curricula that would result in all students becoming science literate.

The education literature was of little help. And so we turned to fields in which there exists a rich literature on design and abundant examples, most particularly (but not exclusively) in architecture and engineering. We soon became aware of a lack of satisfactory language—verbal or visual—for clearly expressing ideas of curriculum structure. And so the project's associate director Andrew Ahlgren and I began to explore possibilities. For at least two years, the walls of my office were covered with an ever-changing array of diagrams and lists, most of them created by Dr. Ahlgren, which invariably caught the attention of staff and visitors, eliciting pointed criticisms and insightful suggestions.

As the concept for *Designs* began to crystallize and draft versions were circulated, two shortcomings were frequently pointed out. One was that the curriculum design process being presented was too far–reaching to be immediately practical; the other that it was too complicated to carry out. The first problem was addressed by working out how educators could make immediate improvements in curricula, while at the same time putting themselves in a strong position to eventually carry out more comprehensive curriculum reforms. The problem of complexity was dealt with not by aggressively simplifying the design process that had been developed—curricula are by their nature complicated—but by seeing how the principles and tools of computer-aided design, so powerful in other contexts, could be applied to curriculum design.

Designs for Science Literacy, like its forebears, is the result of the commitment, ingenuity, and endurance of the entire Project 2061 staff and the contributions of literally hundreds of educators and scientists. I thank them all, and wish especially to acknowledge the extraordinary work of Andrew Ahlgren, my long-standing collaborator.

Given the demand in our country for quick and easy solutions to complex educational problems, it is encouraging that funding agencies are willing to support Project 2061 long enough for works such as *Designs for Science Literacy* to appear. The American Association for the Advancement of Science and all of us who have had a part in this are deeply grateful.

F. James Rutherford
Director Emeritus, Project 2061

Comet Halley, *Photographed from Las Campanas Observatory, Chile, March 1985*

Science Literacy, Curriculum Reform, and this Book

THE BASIC PROPOSITION OF THIS BOOK is that treating curriculum reform as a design problem will contribute significantly to the achievement of the ambitious goals of science literacy. Project 2061 takes science literacy to encompass the natural and social sciences, mathematics and statistics, technology, and their interactions. *Designs for Science Literacy* also pays attention to the need to link these science-oriented studies to the arts and humanities, to vocational education, and to other components of the total curriculum. Although the discussion of curriculum design most often is expressed in terms of that part of the curriculum relevant to science, it will likely have relevance to those other parts as well.

Designs deals with the critical issues involved in assembling sound instructional materials into a coherent K-12 whole. But it does not deal with the development of those instructional materials, and it says little about the practical problems of implementing a new curriculum design in real schools. Instead, *Designs* proposes ways to choose and configure 13 years worth of curriculum materials so that they align with established sets of learning goals, while preserving the American tradition of local responsibility for the curriculum itself.

Whether *Designs'* intent can be achieved will depend on the development of a bank of curriculum materials that have desirable built-in properties, including alignment with specific learning goals, effective styles of instruction, provisions for cognitive and cultural diversity, options for faster learners, and helpful assessment tools.

With the publication of *Designs*, Project 2061 hopes to foster more uniformity among learning goals across the nation, to simultaneously encourage more local diversity in curriculum, and to help launch effective curriculum-reform efforts.

Who Is *Designs for Science Literacy* For?

Designs has been written for five main audiences. Its purpose is to help:

• Administrators and teachers to organize curriculum change in a way consistent with a new national vision of science literacy.

• Developers and publishers of instructional materials to adopt a conceptual framework for the invention and revision of their products, concentrating seriously on the specific learning goals to be achieved.

• Designers of K-12 curricula to consider the science, mathematics, and technology components of the curriculum as a coherent whole.

• Education reform leaders to introduce near-term improvements that will contribute to significant long-term curriculum change.

• College faculty to teach the principles of curriculum analysis and design to new and experienced teachers.

Why Is Curriculum Reform Needed?

Designs presupposes that curriculum reform must be considerably more fundamental and extensive than the tinkering with individual courses and subjects that has been going on for decades. Actually, the call for broad changes in the curriculum has been made over and over again during much of the 20th century. It has reflected certain persistent criticisms of the traditional school curriculum:

The content of the curriculum is not appropriate for meeting the individual and social needs of people living in the modern world. Simply put, the content is obsolescent. Much of what is taught is not needed in everyday life, and much of what is *not* taught *is* needed in everyday life.

The curriculum is a mishmash of topics that lacks coherence across subject-matter domains and grade levels. It is some of this and some of that, with each piece being justified on its own without reference to a conceptual whole.

The curriculum has become grossly overstuffed with topics. The one change that seems easiest to make in a curriculum is to add something to it. So the curriculum's content grows inexorably—often in response to the public demand that schools address social problems, such as alcohol and drug abuse, AIDS, and hazardous driving. New topics are introduced but few disappear. Shallowness is one consequence, incoherence another.

The curriculum does not serve all students equally well. The problem is exacerbated by other factors, such as the inequitable distribution of educational resources and the low expectations held for some categories of students. Rarely is an existing curriculum sensitive or flexible enough to meet the needs of diverse students.

Above all, the curriculum does not produce the learning expected of it. Students may take algebra, history, biology, and the other "right" courses and do well in the course examinations, but extensive research shows they really understand and retain very little of the content. Moreover, development of curriculum and instruction typically takes too little notice of the research on what, how, and when students can learn, depending rather on tradition for topics, methods, and grade-level placement.

Though inadequate teaching plays a part in all of these problems, much of the blame can be laid at the door of the curriculum itself. (Witness the still-to-be-found travesties of trying to teach causes of the seasons in the 2nd grade or electron shells in the 4th grade.)

Designs makes a number of recommendations having to do with unburdening the curriculum. Care must be taken not to interpret these as a call for "watering down." Far from watering down the curriculum, a concentration on understanding key ideas well will enable students to achieve higher standards than those reached by most of the students apparently "doing well" with the current curriculum. The real watering down is quite evident now in classrooms where students receive shallow instruction on so many topics that they retain nothing but a jumble of poorly understood fragments of information.

The Stubbornness of Curriculum

If these long-standing criticisms are valid, why hasn't the curriculum been changed? In spite of many reform attempts, the 20th century has ended with pretty much the same curriculum it began with, plus a heavy sprinkling of new topics. Resistance to change is commonplace in all social systems and institutions, whether they be sports organizations, government agencies, business enterprises, or school districts. Comfort with what is familiar and anxiety about the untried lead teachers, administrators, school boards, state legislators, parents, citizens in general, and even students to be unenthusiastic about curriculum change. For instance, although polls show that parents give low marks to the nation's schools and teachers and support the idea of education reform, they also show that those same parents feel their own schools and teachers are doing well the way they are and do not need a major overhaul.

There is more to the persistence of curriculum than a mistrust of change. A K-12 curriculum is a complex structure that does not stand alone but is an integral part of an even more complex educational and social system, and therefore not easily or simply dealt with. In the United States, the power and resources needed to effect change are widely dispersed, and society is not of a single mind as to what part of the system needs changing or what direction that change should take. As Decker Walker observed in his 1990 book *Fundamentals of Curriculum*, "That the American curriculum influence system can work at all seems improbable, it is so complicated, irrational, disjointed, open, and unpredictable....The entire process can be thought of as a way for the contending parties who share authority for curriculum decisions to negotiate their differences. The parties to the negotiations are the many interested individuals and agencies...playing official, quasi-official, or unofficial roles in the curriculum influence system." What is more, the schooling experiences of both teachers and parents are likely to have been in the traditional curriculum.

Another reason for the staying power of the present curriculum is the lack of obvious alternatives to it. There are a few options for alternative teaching materials and techniques within any given course, but that is about the extent of change usually considered practical.

It is also not clear who will be responsible for designing new curricula: teachers lack the time and resources to do more than make marginal alterations in their own classes, and, in any case, are not trained to be curriculum designers. University faculties in the sciences, for their part, have limited knowledge of how young students respond to subject matter and very few have had experience drafting curriculum for the K-12 grades. Outsiders lack the authority or power to impose change on reluctant school systems.

Bringing about significant and lasting curricular change in the face of this complexity and experience is at best a decades-long undertaking, in spite of the demand by advocates of reform that changes be made in a hurry. If fundamental curriculum improvement is ever to occur, a new process for creating alternatives will have to be developed. *Designs* suggests one such process.

Why Design?

Curriculum is already rich in design: design of lesson plans, design of instructional materials, design of courses, design of course sequences. For the most part, however, these design activities are piecemeal and isolated, seldom greater in scale than a year or two of the curriculum. In other areas of human endeavor (for example, airplane manufacturing, agricultural distribution, or military operations), the design of whole systems has had great benefits—parts work better together, redundancies and gaps are reduced, and less redesign and adjustment are needed.

Fortunately, there are some general principles of how such designing is done effectively. In the belief that general principles of design can have a significant payoff for the quality of the K-12 curriculum as a whole, *Designs for Science Literacy* sketches some design possibilities and calls for practitioners to help fill in the sketch.

Organization of *Designs for Science Literacy*

The Prologue examines some of the basic principles that are useful in almost all forms of design. The eight chapters that follow it are arranged into three parts. Part I, Design and the Curriculum, considers the application of general design principles to curriculum (CHAPTER 1: CURRICULUM DESIGN), and then considers features of curriculum that are most important to design (CHAPTER 2: CURRICULUM SPECIFICATIONS).

Part II, Designing Tomorrow's Curriculum, envisions how curriculum could eventually be designed by selecting from a large pool of high-quality instructional blocks (CHAPTER 3: DESIGN BY ASSEMBLY), describes the desired characteristics of blocks and guidelines for their selection (CHAPTER 4: CURRICULUM BLOCKS), and then imagines what the curriculum-design enterprise in the future may be like for three different school districts (CHAPTER 5: HOW IT COULD BE).

Part III, Improving Today's Curriculum, suggests steps that can be taken to improve an existing curriculum and in the process prepare for its eventual transformation by implementing coherent programs of professional development (CHAPTER 6: BUILDING PROFESSIONAL CAPABILITY), emphasizing understanding of the most important ideas in the currently over-stuffed and shallow curriculum (CHAPTER 7: UNBURDENING THE CURRICULUM), and enhancing the connectedness across subjects and grades (CHAPTER 8: INCREASING CURRICULUM COHERENCE). In the Introduction to Part III, there is an extended passage on practical suggestions for approaching reform, which has relevance for Part II as well.

Designs does not include references to school districts that have made recent progress in redesigning curricula. There are educators who are already doing one part or another of what is proposed, but the design process as a whole is not likely to be found anywhere. Nonetheless, it is only through practitioners—teachers, administrators, materials developers, and curriculum specialists—that the ideas in *Designs* can make sense and lead anywhere. With their help, their experiences can be built into revisions of *Designs* and into future Project 2061 tools for educational reform.

The Epilogue offers another look at *Designs for Science Literacy* and reviews some of the main (and potentially controversial) propositions in the book and attempts to clarify and/or defend them.

Inside the back cover of this book is *Designs on Disk*, a companion CD-ROM. *Designs on Disk* includes a collection of databases, background readings, and utilities to help educators take on many of the curriculum design tasks recommended here. Throughout the book, marginal notes refer the reader to relevant components on the CD-ROM.

The Project 2061 Tool Kit for Education Reform

The curriculum is only one part of a complex education system, and reforming it alone will not suffice to ensure that students achieve science literacy. In the absence of corresponding changes in teacher education, state and local education policies, teaching resources, assessment, administrative practices, and so on, it is unlikely that the curriculum can be changed significantly.

Designs for Science Literacy is meant to be part of a coordinated set of tools that educators can use to improve teaching and learning in science, mathematics, and technology. These tools have been developed as a result of the project's efforts to help reform K-12 education nationwide so that all high-school graduates are science literate. From the start, Project 2061 has defined science literacy broadly to include knowledge and skills in science, technology, and mathematics, along with scientific habits of mind and an understanding of the nature of science and its impact on individuals and its role in society.

Working with panels of scientists, mathematicians, and technologists, Project 2061 set out in 1985 to identify the knowledge and skills that would constitute adult literacy in five subject areas: biological and health sciences; mathematics; physical and information sciences and engineering; social and behavioral sciences; and technology. These learning goals were eventually integrated into the project's landmark document, *Science for All Americans* (1989), which outlines what all students should know and be able to do in science, mathematics, and technology after 13 years of schooling.

In 1993, Project 2061 collaborated with teams of teachers from six carefully selected school districts to create *Benchmarks for Science Literacy*, a curriculum design tool that translates the literacy goals of *Science for All Americans* into expectations of what students should know at the ends of grades 2, 5, 8, and 12. Both documents have had a major impact on education, providing the foundation for national science education standards and helping to shape curriculum frameworks and standards in numerous states and school districts.

Project 2061's tool kit now includes a variety of books, CD-ROMs, and on-line tools to help educators make significant improvements throughout the system:

- *Resources for Science Literacy: Professional Development* (1997) provides educators with valuable background materials to improve their own knowledge and skills.

- *Blueprints for Reform* (1998) outlines changes needed in a dozen areas of the education system to improve learning in science, mathematics, and technology.

- *Dialogue on Early Childhood Science, Mathematics, and Technology Education* (1999) discusses the latest findings on teaching these subjects to preschool children.

- *Middle Grades Mathematics Textbooks: A Benchmarks-Based Evaluation* (2000) and *Middle Grades Science Textbooks: A Benchmarks-Based Evaluation* (2000) present the results of Project 2061's analysis of both widely used and newly developed middle school mathematics and science texts. Similar evaluations of high school algebra and biology textbooks are under way.

- *Resources for Science Literacy: Curriculum Materials Evaluation* (in preparation) reports on Project 2061's approach to analyzing and evaluating instructional materials.

- *Atlas for Science Literacy* (2000) maps out connections among benchmarks to show how student learning progresses over time and how content connects across disciplines.

Eventually these tools and other resources will be merged into a comprehensive, easily accessible on-line system.

Designs is intent upon exploring future possibilities, not about finding immediate solutions to all of our curriculum problems. In that spirit, it is more important that you enter into the conversation than that you agree with what is presented here.

Claude Monet, *The Artist's Garden at Vétheuil,* 1880

PROLOGUE
DESIGN IN GENERAL

The design process is widely used in solving problems and achieving desired ends, even by people with little or no training or interest in the general process. So it makes good sense to begin our exploration of the idea of design with a commonplace example familiar to most people: designing a garden. The garden example enables us to then consider attributes of design in general and the more-or-less sequential set of stages that commonly occur in the process. In Chapter 1, the ideas presented here are applied to the particular case of curriculum design. Chapter 2 considers the basic dimensions of curricula that lend themselves to design.

AN INTRODUCTORY EXAMPLE

Suppose our desired end is to have a backyard garden. One does not have to be an expert to know how to design a home garden, though it is necessary to know something about plants—or about how to find out about them. Our approach need not be orderly, one careful step at a time, but among the things we would surely do are these:

- We would gradually become clearer on what we want from a garden. Will we grow vegetables or flowers or both? Or will we use the garden to hold parties, keep bees, or just enjoy working the earth? And how will we judge the success of our design—will it result in more nutritious meals, honey, reduced florist bills, social prominence, peace of mind, or some combination of those goals? At the same time, we would begin to identify any physical, financial, and legal constraints on what we plan to do.

Design

- To create, plan, or calculate for serving a predetermined end.
- To draw, lay out, or otherwise prepare a design or designs.
- The result of a process of designing.
- The process of selecting the means and contriving the elements, steps, and procedures for what will adequately satisfy some need.
—*Webster's Third International Unabridged Dictionary*

- As our goals and constraints become clear, we would identify some alternative design concepts to help us focus our thinking about design possibilities. A design concept for a garden may be to provide a seasonal succession of vegetables or flowers, imitate an English country garden, attract (or repel) certain wildlife, or simply have a backyard that requires weekends-only maintenance. We would study model gardens in books and magazines, search the Internet, talk to professional gardeners, and look at the gardens of our friends and neighbors—on the chance of finding possibilities that might not have occurred to us.

- We would narrow the possibilities down to a few appealing design ideas that would work within the constraints we face, think over our desired end, and choose an approach that would seem to be the best bet.

- Then we would develop the idea in enough detail to get started actually planning the garden. During this stage, trade-offs would have to be considered—a choice, for instance, between the desire for large shade trees and for sun-loving plants. Our plan would very likely be in the form of a sketch showing how the plants and other features of the garden would be placed. In developing the final design, we could call on experts for advice, or use a commercially available computer-assisted garden design program.

- As the actual work in the backyard progressed, we would come up against unexpected difficulties, forcing us to modify the original design—or even to choose an alternative design concept altogether.

- Even with the garden in place, the design challenge would not be over, as any home gardener knows. Maybe the actual garden would not look exactly like the design, or it would be just like it but would not please us. Maybe mistakes made in implementing the design would now show up. In other words, we would discover or decide that modifications were needed. We would have to make allowances for the fact that even if the garden were entirely satisfactory at first, it could turn out that as the plants matured, the relations among them would change enough to require still other modifications.

As our discussion of the garden example shows, there is nothing mysterious about design—architectural, engineering, horticultural, or any other kind. But as straightforward as it is, design does have certain features that people sometimes overlook in solving problems and achieving desired ends. It is worth, therefore, exploring the design process more generally.

ATTRIBUTES OF DESIGN

The story above would have been much the same had the desired end been the Brooklyn Bridge rather than a home garden. Each particular design undertaking has its own special features depending on traditions and circumstances, but in general the process applies equally to the design of any object, process, or system. As our garden example illustrates, design has these attributes:

Design is purposeful. The purpose (desired end) may be to improve a curriculum's effectiveness, solve a problem of traffic congestion, replace pesticides with crop diversification or rote learning with understanding, exploit some existing technology in new ways, or create a new product or service. In practice, there may be many purposes in a design undertaking. They may be in harmony or in conflict, explicit or hidden, immediate or long-range, political as well as technical. Whatever the mix, the designers of a project are better off if they know all of the purposes at the outset, so that they can respond accordingly.

Design is deliberate. The Brooklyn Bridge did not just appear one day in all its glory, nor did it evolve over the decades from a pontoon bridge, nor did it result from a lot of workers showing up each day and deciding what to do next. That may seem obvious when thinking of a bridge, but not always when thinking of a school curriculum.

Leonardo da Vinci's diagram of ribbed wing for a flying machine.

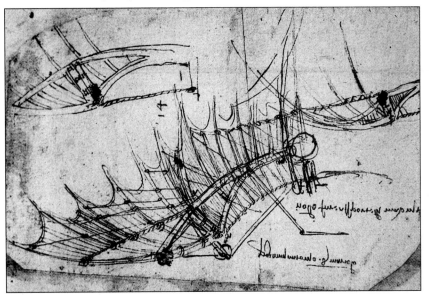

For an engrossing account of the importance of creativity in design, read the story of the invention of the chronometer in *Longitude: The True Story of a Lone Genius Who Solved the Greatest Scientific Problem of His Time*, by Dava Sobel (1995)

Design is a conscious and deliberate effort to plan something—an object, event, process, or system. A curriculum is one such thing and just as susceptible to deliberate planning as a garden, ship, weapon, banquet, traffic pattern, ballet, assembly line, or scientific experiment.

Design is creative. For all its practical focus, design is not some mechanical process that invariably leads to success. Like science itself, which is often misrepresented as a fixed sequence of steps ("the scientific method"), the design process is a highly variable and creative process. Nevertheless, like scientific inquiry, it has certain features that show up again and again. At every stage of a design undertaking, whether it happens to be the design of a new hospital or a telephone routing system, there are opportunities for innovative thinking, novel concepts, and invention to be introduced. Perhaps it is fair to say that design is powerful precisely because it is at once systematic and creative, feet-on-the-ground and head-in-the-clouds.

Design operates on many levels. There was a design for the Brooklyn Bridge as a whole. But there also had to be a design for each of its parts and construction operations. Those designs covered such activities as cutting and transporting the stone for the towers, fabricating the cables, laying the roadway, and even creating special tools. An essential design principle is that the design decisions at one level must be compatible with those at the higher levels. (In the case of the Brooklyn Bridge, a design for a rivet that did not match the expected tension on a main girder could eventually have led to the collapse of the whole bridge.) Many common engineering tasks involve selecting among parts that already exist, so what parts are available may affect the larger-scale design. In other situations, the larger-scale design may also require the design of special new parts and processes.

Design requires compromise. Design is not the pursuit of truth or perfection, but it has to get the job done. Architects and other designers are expected to come up with practical solutions that work well enough in the circumstances. In reaching decisions leading to such practical solutions, trade-offs are made among benefits, costs, constraints, and risks. Therefore, no matter how careful the planning or how inventive the thinking, designs always end up having shortcomings when viewed from one perspective or another, but they are knowingly accepted as reasonable compromises. Some shortcomings, however, may be unanticipated and may not show up in the design process until the designed object or process is put into use, so the design should include provisions for keeping an eye on its success.

Design can fail. There are many reasons why designs may fail. Stereoscopic movies were a marketing failure, even though technically they worked as planned. Leonardo da Vinci's flapping-wing aircraft didn't work for mechanical and conceptual reasons (because no adequate source of power was invented until 300 years later, and because in any case he based his design on the wrong model—birds). Hydrogen-filled dirigibles failed because they were unsafe, as dramatically demonstrated by the *Hindenburg* disaster. A designed system can fail because one or more components fail or because the components do not work well together, even though each works well enough by itself. In most cases, however, designs are neither wholly satisfactory nor abject failures, and so a key element in design is the provision for continuing correction, assessment, and improvement, both in the initial design process and after.

"There is no perfect design. Accommodating one constraint well can often lead to conflict with others. For example, the lightest material may not be the strongest, or the most efficient shape may not be the safest or the most aesthetically pleasing. Therefore every design problem lends itself to many alternative solutions, depending on what values people place on the various constraints. For example, is strength more desirable than lightness, and is appearance more important than safety? The task is to arrive at a design that reasonably balances the many trade-offs, with the understanding that no single design is ever simultaneously the safest, the most reliable, the most efficient, the most inexpensive, and so on."

—*Science for All Americans,* p. 28

A 1950's design for 3-D movies did not survive.

Another version of design stages is outlined in *Design and Problem Solving in Technology* by John Hutchinson and John Karsnitz (1994): (1) Identifying problems and opportunities; (2) Framing a design brief; (3) Investigating and researching; (4) Generating alternative solutions; (5) Choosing a solution; (6) Developmental work; (7) Modeling and prototyping; (8) Testing and evaluating; and (9) Redesigning and improving.

Design has stages. Design is a systematic way of going about planning. While it does not consist of some inflexible set of steps to be followed in strict order, design does involve certain stages that take place at one time or another in the process, or at several times in the process. Practical design is likely to include looping back between stages. Decisions made at one stage may require reconsideration of how they interact with decisions made at other stages. As a result, a design almost inevitably evolves somewhat in the process. The original desired end itself is likely to be clarified as the design progresses, and it may even have to be modified as constraints and costs are discovered. In brief, design involves the following four stages:

- Getting as clear as possible in the design specifications: precisely what is to be achieved and what constraints must be accepted.
- Conceptualizing several alternative design possibilities and thinking about each enough to be able to choose one as a best bet to develop further.
- Developing a complete design, testing aspects of the emerging design along the way, and making adjustments as needed. This stage almost always requires trade-offs involving goals, constraints, benefits, costs, risks, and desirable design features.
- Refining the complete designed product on the basis of experience and feedback from users.

The remaining sections of this chapter elaborate on these four stages.

Establishing Design Specifications

Being purposeful, design is an effort to achieve something specific—devise a means to reach a desired end, satisfy a need, or take advantage of an opportunity. Thus, one of the first steps in design is becoming clear on just what is to be achieved. At the same time, consideration must be given to the essential fact that design is always confronted with constraints on what it can do—time and money limitations, legal restrictions, political considerations, cultural traditions, the laws of nature, and more. Together, goals and constraints determine the specifications a design is expected to meet.

Moreover, design specifications can be technical, aesthetic, financial, political, or moral. The design requirement for the Boeing 777 that it be able to reach an airport safely if one of its two engines fails while over the ocean was both technical and moral. Design specifications can be attributes desired by the client or designer, or they can be demands imposed on them from the outside. In fact, much of the work in developing a design is figuring out how to respond to—or get around—constraints while still reaching the desired goals and serving the purposes that initiated the design undertaking in the first place.

Goals

Purposes are usually couched in sweeping language, such as to create a health-care delivery system that is more cost-effective than current ones, an electric vehicle that can travel long distances without recharging, a beautiful backyard garden, or a K-12 curriculum that enables all students to become science literate. Such general purposes then have to be transformed into more concrete goals. Design can get under way before all of the goals have been clearly defined, but there must be more specificity than is provided by the usual statement of purpose. Once the design process is started, the clarification of the goals continues, and they become more and more specific. Occasionally new goals may be added, but for the most part goals are progressively derived as expressions of higher-order goals.

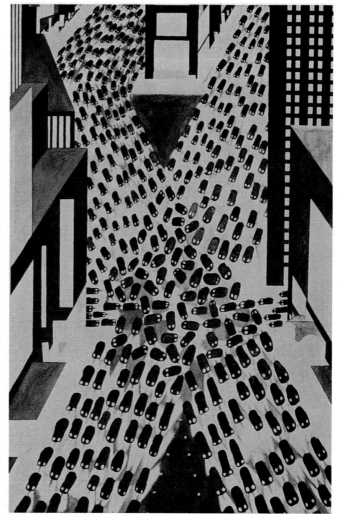

It is important to establish just how specific the goals have to be. Suppose, for example, the mayor of New York City orders city officials to come up with a way to speed up crosstown (east-west) automobile traffic. In response, the officials could try one thing or another and see if crosstown traffic speeds up. But that could easily make matters worse, and so, before tackling the problem, traffic-pattern designers would want to know more specifically what they are expected to achieve in "speeding up crosstown traffic." Are all crosstown streets to be included, or only the main ones? Are all the boroughs to be speeded up, or only Manhattan? Does it mean crosstown traffic all day long or only at certain peak periods? How much faster will do—10 percent? Are we talking about an equal increase for each street or an average increase for all streets? And so on.

Setting goal specifications is rarely as simple as it may seem, however. Often, as goals become clearer, they evoke tensions among groups that do not share the same interests or beliefs. Scientists may be in accord, for instance, with the general proposition that the

United States should create and sustain a vigorous program of space exploration. Yet if Congress translates that idea into increased funding for a space station, scientists who believe that unmanned space exploration is more productive may rise up in opposition, as may scientists in other fields who see their own funding put in jeopardy. Or speeding up crosstown traffic may reasonably be expected to slow down traffic going in other directions. To get one thing, we decide to—or have to—sacrifice another, or to get more of something, we agree to settle for less of something else.

And goals run into constraints. In fact, some goals can become constraints on others. The mayor's dictum could be read as "Speed up traffic in one direction in a way that doesn't slow down traffic in the other direction." The mayor may also stipulate other requirements, such as that the new traffic-flow design must not put citizens and institutions at risk, increase city expenses or decrease city revenues, make it difficult for deliveries to be made to stores, or otherwise impede businesses. Somewhere in the design process, something will have to give, for it simply may not be possible to create a design that fulfills the mayor's purposes and meets all of the conditions he has imposed. However, before informed trade-offs can be made, the constraints and goals have to be specified more precisely.

One can expect goals to be modified in the design process as knowledge of the situation grows and as constraints appear. If, for instance, traffic-flow studies revealed that crosstown traffic is unacceptably slow on only a third of the downtown streets, it may be possible to redefine the goal to bring those streets up to speed without improving the others. Whether this redefinition of the goal would be acceptable to the mayor may have more to do with political considerations than with strictly technical ones.

There is often more to goals than what appears on the surface. In the traffic case, the fictional mayor's real purpose may be to increase his chances of reelection by showing a readiness to deal with a long-standing city problem and to deflect attention from his lackluster performance on other city problems, such as housing and crime. Our real goal in creating a backyard garden may be to keep up with the Joneses or to provide a worthwhile activity for a retired spouse, goals that may not be met if they are made too evident. To take an example from the schools, a goal such as reducing truancy, usually expressed in educational terms, also has as a silent partner—a community goal of keeping unsupervised young people off the street during the day. Designers need to be aware of the possibility that success requires taking into account some goals that have not or cannot be made public.

Constraints

Clarifying constraints is as much a part of design as delineating goals—the goals saying what is to be accomplished, the constraints specifying the limits. In a sense, much of what is said above about goals applies to constraints. For instance, at the beginning of the design process, constraints to be dealt with are often expressed in general terms, such as "must not exclude handicapped customers," "be able to use existing runways in international airports," "be environmentally benign," or "do not increase instructional costs." None of these particular limiting conditions are stated well enough for design purposes, and so they must be transformed into specifics. For instance, the runway constraint will need early translation into specifications limiting the total permissible weight, brake performance, and maximum take-off distance of the aircraft based on knowledge of the physical properties of runways at international airports.

Some of the most stringent limits on a design are those imposed from the outside. For example, a new museum must meet all of the building codes in the community where it will be located; a new jet aircraft must comply with the safety requirements of the Federal Aviation Administration; a new medicine must meet the effectiveness and safety demands of the Food and Drug Administration. Interestingly, not all such external requirements emanate from government agencies or have the authority of law. Professional and trade associations set expectations and sometimes explicit standards that influence or limit design possibilities.

Just as goals may be modified during the design process because they conflict with one another, so too may constraints. As constraints become more and more precisely defined, it sometimes becomes clear that

"Every engineering design operates within constraints that must be identified and taken into account. One type of constraint is absolute—for example, physical laws such as the conservation of energy or physical properties such as limits of flexibility, electrical conductivity, and friction. Other types have some flexibility: economic (only so much money is available for this purpose), political (local, state, and national regulations), social (public opposition), ecological (likely disruption of the natural environment), and ethical (disadvantages to some people, risk to subsequent generations). An optimum design takes into account all the constraints and strikes some reasonable compromise among them."
—*Science for All Americans, p. 28*

they block all possibilities for achieving the stated goals. This situation can lead to a renegotiation to eliminate or modify some of the constraints. That may sometimes be difficult or even impossible. With regard to the runway requirement, for example, the manufacturer may try to persuade the authorities that technological innovations like a new kind of landing gear justify raising the maximum permissible landing weight. On the other hand, constraints derived from the laws of physics such as the force of gravity at the earth's surface, are simply not negotiable.

In short, just as goals may turn out to be so unrealistic that they have to be abridged in the design process, constraints also may be so paralyzing that they are challenged along the way: new materials, technologies, and processes can be invented, legal regulations can be overturned, public opinion can be molded, expectations can be transformed, funding priorities can be changed, and so forth.

CONCEPTUALIZING A DESIGN

Behind every interesting design—the Panama Canal, the *Whole Earth Catalog*, or the Cannes Film Festival—there is an interesting idea, or several interesting ideas, often likely to be sketchy, rarely precise. Such ideas—"design concepts"—come before designs, and although they rarely survive the design process intact, they are essential for getting started.

The mythical griffin blends aspects of different design concepts.

A design concept is any overarching idea, or set of ideas, that suggests the character of the thing to be designed. The designers of the United Nations Headquarters complex in New York City, for instance, set out to create a facility that would proclaim the dignity and significance of the infant organization, yet serve as a practical "workshop for peace." This metaphorical workshop for peace would be international in spirit but still live in harmony with its surroundings, and would point to the future rather than recall the past. Observers differ about how well the final design embodies those concepts, and surely those same guiding concepts might well have led to other designs. But for the designers themselves, the concepts provided a powerful unifying theme. (Design concepts can, incidentally, also play an important role in generating enthusiasm and funding for the project.)

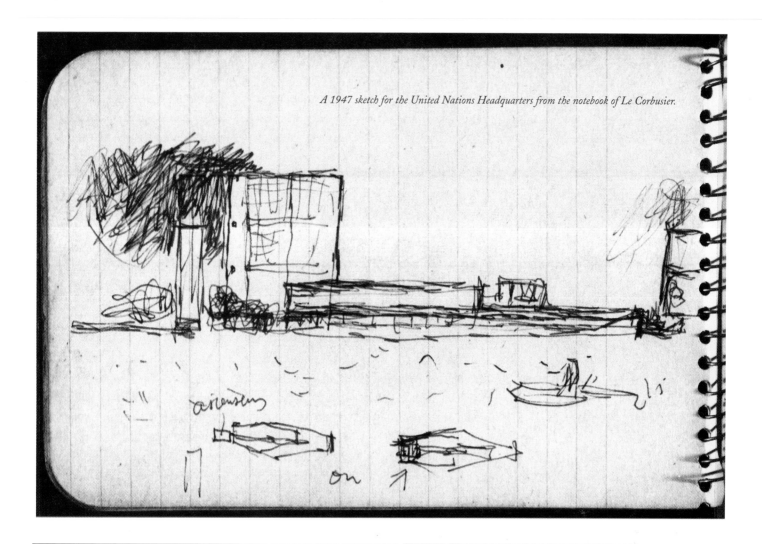

A 1947 sketch for the United Nations Headquarters from the notebook of Le Corbusier.

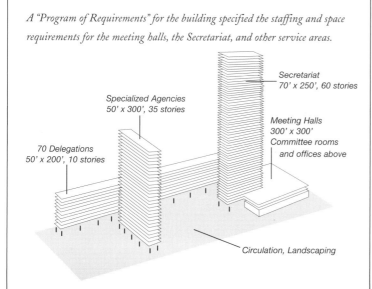

A "Program of Requirements" for the building specified the staffing and space requirements for the meeting halls, the Secretariat, and other service areas.

Secretariat
70' x 250', 60 stories

Specialized Agencies
50' x 300', 35 stories

Meeting Halls
300' x 300'
Committee rooms
and offices above

70 Delegations
50' x 200', 10 stories

Circulation, Landscaping

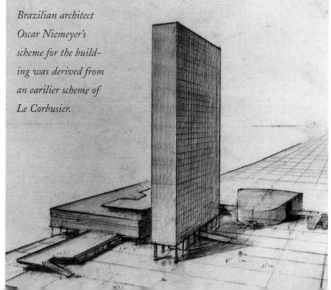

Brazilian architect Oscar Niemeyer's scheme for the building was derived from an earilier scheme of Le Corbusier.

For an interesting account of different design concepts applied to a major undertaking, see *The Path Between the Seas: the Creation of the Panama Canal, 1870-1914* by David McCullough (1978).

Often, the design concept can be captured in a visual sketch, but sometimes it is expressed in prose or a combination of words and images. In more mundane situations, however, there are existing designs so well worked out and widely known that we can just adopt one (a station wagon, a Cape Cod house, a college-preparatory curriculum) or adapt one (a station wagon with bucket seats, a Cape Cod house with a solarium, a college-preparatory curriculum with community projects). Sometimes there is no attractive precedent that will serve as a design concept, and instead the designers emphasize some aspect of an already developing design—a purpose, component, feature, or effect—to provide a character for the design and guide its further development.

In design, there are always alternatives. There is rarely just one right way to do things. Generating and considering competing possibilities is a fundamental step in successful design. Choosing among alternative design concepts may not be easy, but the need to consider and make conscious choices forces designers to think on a grand scale before turning to the details.

How to decide among possible design concepts? Formal analysis of the relative benefits and limitations, costs, dependence on other systems, and possible side effects may help, but intuition based on experience and knowledge of the territory also come into play. Naval architects know a lot about ships, movie directors about cinema, and so forth. But just as the consideration of alternative design concepts is important, it is equally important that design not stall out at the stage of considering alternatives. It is simply too costly to develop many competing designs simultaneously.

DEVELOPING A DESIGN

Once progress has been made toward setting goals, identifying constraints, and selecting a design concept, the main task of developing a full-fledged design can proceed. For a monumental example, consider the Brooklyn Bridge. The goal was to find a way of moving large numbers of people back and forth across the East River between Brooklyn and Manhattan. That led to a design concept of a two-tower suspension bridge (rather than any other kind of bridge, or a tunnel or more ferry boats) high enough for oceangoing ships to pass underneath. Then the specifications for every feature of the bridge were formulated—where the towers would be located, what they would be made of, how high they would be, what the wire cable would be made of and how it would be installed and anchored, what the dimensions and slope of the spans would have to be, and so on.

In developing a design for a product or system, new opportunities and ideas may emerge, along with the inevitable unanticipated impediments. Not all of the ideas can be exploited and not all of the impediments can be overcome, so choices have to be made *as the design effort proceeds*. Estimating relative benefits, costs, and risks provides a basis for making trade-offs among the possibilities. For example, construction of the towers of the Brooklyn Bridge involved so many worker injuries and deaths that the chief engineer halted construction at a depth considerably less than that called for in the design; he felt that saving the workers' lives outweighed the perpetual risk of the bridge one day collapsing. So far, the trade-off has been successful; the Brooklyn Bridge is now over a century old. However, there are extreme cases of designs in which none of the trade-offs is desirable, acceptable, or even tolerable, and it may be necessary to reject the selected design concept and take up another one for development.

Some Helpful Strategies

Actually developing a design can be rather easy—or it can be daunting; it depends on the complexity of the challenge. To design our garden may take only a few days or weeks and require little help, whereas designing a space station takes years and involves a cast of thousands. Most design challenges—including curriculum—fall between those two extremes of scale. In even moderately complicated design undertakings, there are some strategies that can help to deal with the complexity. One of them is to copy or modify an existing design; a second is to divide the design task into component parts that are individually more manageable than the whole thing; and a third is to plan on testing the maturing design repeatedly during design development. Following is a brief look at these three strategies.

Copying or modifying an existing design innovation. In turning to an existing design for guidance, the presumption is that it represents a successful design. Because the actual Brooklyn Bridge did work (in the sense of doing what was expected of it and not falling down), many other bridges have used very similar designs. However, design failures can occur when modifications of basically successful designs are carried too far. In his 1994 book *Design Paradigms: Case Histories of Error and Judgment in Engineering*, Henry Petroski claims that the history of bridge building is littered with the stories of designs that have been extrapolated too far from known successes.

Since it is rarely possible to copy an existing design down to the last detail in new circumstances, modifications are usually necessary, although they may entail risk of not working as well. But then there is usually an even greater risk in not making modifica-

The size and shape of the early Volkswagen is recaptured in a fresh 1998 design.

tions when fitting a given design, no matter how successful, to a different situation. In adapting an existing design, designers must call upon accurate information, relevant experience, and known principles—which may not always be available. And they must be alert to the fact that parts from different designs may not work well together.

When they are known, established principles about what works can be extremely important. To create a design for spanning a 25 percent wider river, it is not simply a matter of planning to make the bridge 25 percent longer. Rather, the new design must be guided by physical principles that relate the strength of beams to their length and cross section, or that relate the wind drag on a cable to how high it is off the ground. Consider taking a curriculum design that has been found to be successful in middle-class suburbs and modifying it for inner-city schools: How much more confidently that could be done if there were known principles for how students in those two types of schools differ in how they can best learn. Curriculum design is often limited because the underlying principles about teaching and learning are not known well.

Copying an existing design could involve identifying a completed, operating object or system (the Paris Opera House, Yale University, the United States Constitution, the Pony Express) and studying it to derive its design—that is, describe it in a way that makes possible fashioning another like it. If, for example, a school district's operating K-12 instructional program were to be copied, it would have to be studied and described to be able to provide a curriculum design for another school district.

Compartmentalizing the design components. Development of a complex design can be greatly simplified if it can be divided into parts, each of which then becomes a separate design challenge. Suppose the concept we adopted for our backyard garden called for an area dedicated to easy-care perennials and an area dedicated to vegetables for the family, and that the borders of each had been set. It would then be easy to concentrate on the design of each component separately, perhaps with each being handled by different family members. When Boeing designs a new aircraft, it typically contracts with other companies to design and manufacture subsystems (for power, navigation, communications, etc.) after writing the specifications (goals and constraints) for each. And it may very well plan to use some "off the shelf" parts in its design, rather than designing new ones. The specs for each subsystem, of course, must include relationships to

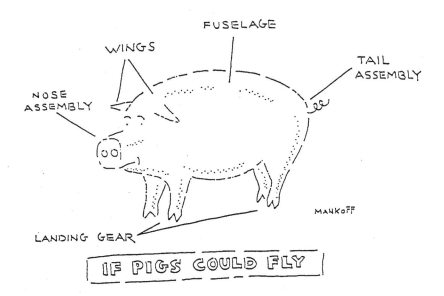

other parts. For example, all electrical components will likely be expected to run on the same voltage. If the specifications for one part have neglected the requirements of other parts it must connect to, it may work poorly or not at all. In the end, however, Boeing is responsible for making sure the whole design works—which is not a foregone conclusion just because all the subsystems individually pass muster.

Although curriculum design is addressed at length later, it provides an excellent example right here for mismatched parts. To the degree that curriculum design for the whole K-12 range is done at all, the task is usually divided into nearly independent parts. Sometimes the division is by subject-matter domain (the reading, mathematics, science, history curricula, etc.), sometimes by grade level (elementary, middle, and high school, or even grade by grade), sometimes by track (vocational, general, college preparatory, advanced placement), and sometimes by combinations of these (the college-prep foreign-language curriculum). The trouble seems to be that, whatever the quality of the design for each component, the parts usually do not get put back together to form a coherent whole that optimizes students' learning over their whole K-12 range of instruction. Good curriculum design should attempt to optimize learning across the entire curriculum, not just unit by unit, subject by subject, or grade by grade.

"Large changes in scale typically are accompanied by changes in the kind of phenomena that occur.... Buildings, animals, and social organizations cannot be made significantly larger or smaller without experiencing fundamental changes in their structure or behavior."
—*Science for All Americans*, pp. 179-80

The Tacoma Narrows Bridge provides a famous example of the terrible consequences that can result from ignoring a known principle. Designers of the unprecedentedly long span correctly extrapolated the increased stiffness required to control up-and-down vibration, but neglected to consider the twisting oscillations that were previously unimportant; when built in 1940, the bridge disintegrated in the first high wind.

"Designs almost always require testing, especially when the design is unusual or complicated, when the final product or process is likely to be expensive or dangerous, or when failure has a very high cost. Performance tests of a design may be conducted by using complete products, but doing so may be prohibitively difficult or expensive. So testing is often done by using small-scale...simulations...or testing of separate components only."

—*Science for All Americans*, p. 29

Testing in the development stage. No matter how good the existing design model is, how thoughtfully the design challenge is subdivided, and what sound principles and storehouses of information are drawn upon, the process of designing a complex system sooner or later (usually sooner) is beset with uncertainties. Designers want to know if they are on the right track before they get irrevocably committed to a design concept, and then they need to find out whether the various components really will perform as required by the design specifications. They can accomplish this in the development stage of design by testing components of the maturing design frequently. Before building a new middle-school curriculum around parent volunteers, for example, educators should see how many volunteers could be turned up in the community. Or before basing a self-paced mathematics curriculum on computer tutorials, educators should test how local students learn from tutorial software.

The best way to test the component of a design is to make a prototype of it and see how well it does what it is supposed to. In some cases, this can be done by using small-scale trials. In designing new aircraft, for instance, scaled-down wing shapes can be

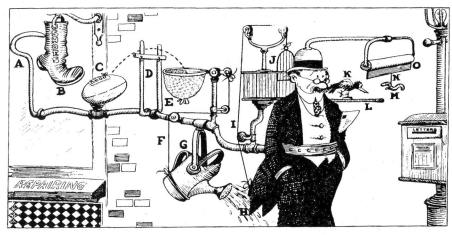

"Keep you from forgetting to mail your wife's letter"—Rube Goldberg™

tested in a wind tunnel; in designing new skyscrapers, computer simulations can be used to answer "what if" questions about the effects of wind shear on a building of various dimensions and orientations. In other cases, it is possible to test a process on a small sample. For instance, aspects of the design for the next round of the national ten-year census of the entire population are tested on a representative sample of some hundreds of households to see how people will respond to the questions. Testing all of the components of a design may not be necessary or even feasible, but the practice is to test those components for which the greatest uncertainty exists or that are most crucial to the success of the design. The question remains open, of course, as to whether the components will all work together when the time comes, and thus the design as a whole will eventually have to be tested. In a home-entertainment system, some top-rated amplifiers may not work well with some of the top-rated loudspeakers. In the school example, the volunteer parents, however numerous, may or may not be effective coaches for the computer tutorials that worked well under experienced teachers.

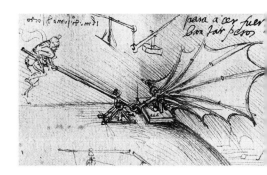

Testing the lifting power of a wing.

Design Decisions

Ideally, one would like to be able to create objects, processes, or systems that would perfectly serve all the identified goals and do so at low cost and without any risk. Alas, needed resources are not always available; insurmountable constraints get in the way, things cost too much, and, like it or not, there is always risk. Because the design process is the pursuit of acceptable solutions, compromises are expected between what is desired and what is feasible. Making design decisions often comes down to agreeing on trade-offs among desired features on the basis of estimates of their relative benefits, costs, risks, and the associated trade-offs.

Benefits and costs. It is natural for the proponents of a new design for a product or process to emphasize its possible benefits: the disease it will cure, the faster it will get people from one place to another, the greater grain yields that will result, the more that students will learn in a year. It may not be known until sometime after the product exists or the process is in effect whether those benefits actually will accrue—although for some few things, such as medical drugs and procedures, stringent testing is required before they can be put on the market.

But potential users are likely to respond to projected costs as well as to claimed benefits. Hence, cost-effectiveness is not far from the minds of designers in any field. The following questions suggest that there are often social costs, as well as immediate and long-term financial costs, to consider:

- Who are the main beneficiaries of the proposed design? Who will receive few or no benefits? Will people other than the beneficiaries have to bear the costs? Who will suffer if the design is implemented? Who will suffer if it is not? How long will the benefits last? Will the design have other applications?
- What will the proposed design cost to build and operate? How does that compare to the cost of alternatives?

These questions are based on similar ones posed in the section on technological decision making in Chapter 3 of *Science for All Americans.*

"The best laid schemes o' mice an' men
Gang aft agley,
An' lea'e us nought but grief an' pain
For promised joy."
—Robert Burns, *To a Mouse*

• What people, materials, tools, and know-how will be needed to build, install, and operate the proposed new system? Are they available? If not, how will they be obtained, from where, and at what cost? What energy sources will be needed for construction or manufacture, and also for operation, and at what cost? What will it cost to maintain, update, and repair the design if it is implemented?

Risk. Seldom are all the effects of a design reliably predictable. This means, of course, that we can never count on getting all the benefits that we hoped for from a design, although it happens sometimes that totally unexpected benefits arise to surpass those intended by the design. But also weighing heavily on the mind of the designer is the specter of unwanted outcomes. There is always risk—the issue is never risk versus no risk—and so statisticians and engineers have worked hard to develop reliable ways of estimating risk. A formal analysis of risk involves estimating a probability of occurrence for every undesirable outcome that can be foreseen and also estimating a measure of the harm that would be done if that outcome did occur. The risk of each undesirable outcome is the product of its probability and its measure of harm. The sum of these risk estimates (perhaps with adjustment for correlation among them) then constitutes the total risk of the design. But such elaborate theoretical risk estimates are always difficult and sometimes impossible to make. Instead, it is usually necessary to settle for rough estimates of how the risks associated with a proposed design compare to those of other possible designs, including the design it is intended to replace (which, however well established, is likely to carry risks of its own). Whatever the outcome, it is still the case that the risk associated with a particular design can never be reduced to zero, and so designers take steps to minimize risk at reasonable cost.

One hedge against failure is what is called overdesign—for example, making something stronger or bigger than is likely to be necessary. Another hedge is redundancy—building in one or more backup systems to take over in case the primary one fails. On NASA missions, the onboard backup computer also has a backup. In an education context, some provision for remediation should always be available if the first-line instruction does not succeed.

This vintage, finned automobile is an example of carrying one design characteristic too far.

If failure of a system would have very costly consequences, the system may be designed so that its most likely way of failing would do the least harm. This requires a value judgment of the kind, "If there is a possibility of some failures, let's try to err on the side of bad outcome X rather than the even worse outcome Y." One example of "fail-safe" design is the on/off switch of an electric lawn mower—if the switch breaks, better it should get stuck in the off position than in the on position. Another example is the U.S. legal policy under which uncertainty about guilt in criminal cases leads to acquittal rather than to conviction (on the value judgment that it is better to free the guilty than to punish the innocent). Of course, not everyone may agree just what the risks and the costlier consequences are. For example, debate persists in education policy about whether the risks of promoting students with dubious achievement should be preferred to the risks of holding them back. Currently it is most common to see greater risk in holding marginal students back than in promoting them, and so there is a preference to err on the side of being too optimistic rather than too pessimistic (a literally "fail-safe" policy). Also, as suggested earlier, the likelihood of failure in either direction is reduced by doing more testing to develop a more robust design.

Trade-offs. The term "trade-off" has come into vogue only in recent years, but the idea behind it is venerable. It points to the age-old practice in human affairs of making compromises—of sacrificing in one way to gain in another. A trade-off is based on the common-sense notion that only very rarely can one have everything one wants—espe-

"An Environmental Consideration"

cially if others are also to get what they want. Trade-offs are solutions, and they result from such things as people needing to share a fixed total amount of some available resource, such as money or time or raw materials, or from their being in conflict over incompatible values, as in exercising both generosity and frugality, or in their wanting to treat all people equally and yet respond to individual needs. Trade-offs also occur because individuals and groups do not all see things the same way. When they are at odds with regard to something in particular, each party may have to give ground to come up with something that is agreeable (or at least tolerable) to all parties.

The practice of making trade-offs is central in design. Trade-offs can be used to settle differences in goals (for example, give up the shade trees, or some of the shade trees, in the backyard garden design, to ensure having more sunlight for the flower beds and vegetable plot); negotiate constraints (a downtown office building design may be permitted to be higher than the code stipulates in return for using less of the ground area than is the rule); or to balance goals against constraints (the mayor will settle for less improvement in crosstown traffic than he would like so that north-south traffic will not be slowed down).

Making trade-offs in these ways can be simple or complicated depending on what the design undertaking is. In many situations, it may not be possible to base design trade-offs on rigorous benefit-cost-risk analyses because sufficient information is not available, applicable principles are not known, or the complexity is simply too great. In other situations, it may not make sense to back every trade-off with such analysis, if for no other reason than that the cost in time and money would become prohibitive. Good judgment needs to prevail in this as in other aspects of design. However, there is little doubt that good design is fostered by (1) making trade-offs deliberately and (2) doing so in the light of what is known about benefits and costs and risks. Benefit-cost-risk analysis requires some way of assigning relative magnitudes to benefits, of quantifying costs, and of estimating risks. CHAPTER 1: CURRICULUM DESIGN includes a discussion of some of the difficulties of doing such analysis in an education context.

REFINING THE DESIGNED PRODUCT

A design is not the product itself. Indeed, the thing itself—the actual garden, canal, jet aircraft, chair, dress, building, curriculum—often does not perfectly match a design. The difference may result from some flaw in the logic of the design, poor information on which the design is based, or the unexpected influence of factors not

The time given to the design process itself is also subject to trade-offs. At some point, it may seem better to go ahead with an imperfect product rather than to wait until the design is improved further.

covered by the design. If discrepancies become too great, the production may be aborted and, in a now-famous phrase, sent "back to the drawing board." On top of that, designs can usually be interpreted in different ways, and hence what the client believes the design intends may not perfectly match what the designer intends.

When components of a product are tested separately before they are implemented together, feedback and refining have already begun. But no matter how closely the product matches the design, and no matter how well the designed components performed when tested during development, no designed product turns out to work exactly as intended when it is all put together and used in the real world. Even products that are outstanding at first often become less so as new demands are placed on them—the air traffic control system that was designed to track a certain number of planes but that now must track many times that number is a current example. Other products continue to perform well enough, but new technology and policies may make them obsolescent. Anyone who has bought a computer recently can attest to that phenomenon. For all these reasons and others, design is never complete.

Another reason that designs need to be refined even after they become actual things is that they are used or affected by human beings. Assumptions are made, often implicitly, about the interaction of the product and people—that all pilots, not just test pilots, will be able to use the controls properly; that drivers will obey the signal lights; that office workers will receive training in the use of the new computer system; that we will tend our new garden faithfully; and so on. In many cases, what later gets labeled as "human error" may be the result of making naive or unwarranted assumptions about how people will actually use the designed product.

"Well, back to the old drawing board."

Given all these ways in which a finished product can fall short of expectations, good design practice makes provisions for systematic retesting. Performance feedback on a design may come from instruments (say, for monitoring stresses on a bridge), from direct observation of operations, or from testimony from users. In the case of air traffic control, inspectors can measure traffic delays, look over the shoulders of operators and pilots, or interview them. In the case of curriculum, supervisors can observe classrooms or interview teachers or students.

Feedback can also come from assessment of whether the initial goals for the design are being met and the constraints adhered to. In air traffic control, do annual records show that traffic flow is sustained with a tolerable number of accidents? In curriculum, do student achievement records indicate at least short-term learning of concepts and skills—and better, do subsequent studies of graduates show long-term retention? Periodic assessments can identify unanticipated shortcomings and sometimes lead to suggestions for modifications. Because of the incremental changes made in its design based on feedback from users, the Boeing 777 that rolls off the assembly line today is not identical in design to the first one that emerged. The design, under which additional products will be produced, is progressively modified on the basis of real-world experience with the early products. On the other hand, when there is a single, unique product—say, the United Nations Headquarters—fix-ups have to be made in the actual product. The initial *design* followed in producing it is not likely to be modified—on paper—unless similar products are planned. One could see the changes in a product as implying a modified "design" in a descriptive sense. Sometimes, but not often, feedback from actual use is so negative that it becomes necessary to rethink the entire design. But in most cases, designs just replace a piece at a time with new and/or better ones.

Since our garden will not turn out to look exactly like our design for it, why bother with design at all? We could, instead, just start planting things at random in the backyard and eventually a garden would exist—perhaps even a beautiful one. But the odds are very much against it, especially if we expect the different parts of the garden to relate harmoniously to one another. Even though there is some difference between our actual product and the design on which it is based, we are more likely to be satisfied with the product than if we simply forged ahead without careful planning. Obviously, that is even more true for more complicated systems having criteria for success that go beyond pleasing us subjectively.

LOOKING AHEAD

Design does not automatically and inevitably result in successful products. Nor has everything admirable that exists necessarily been designed. But among those things that can be and ought to be designed as whole systems are curricula. In this Prologue, we have given little attention to how design relates to curriculum—in order to emphasize what is common in almost all design in the world outside education. The first chapter in what follows considers how the principles in this Prologue can be applied to the particular case of curriculum, and the chapter after that elaborates on what properties of curriculum are appropriate to design.

What the state framework specified

What the curriculum committee designed

What the superintendent requested

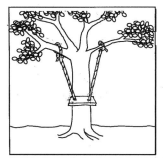

What the board of education approved

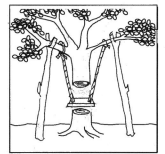

What the publisher actually produced

What the literacy goals had intended

PART I
DESIGN AND THE CURRICULUM

Consider, to begin with, the traditional method we use to go about creating a new curriculum:

> We start with what the existing curriculum is, rather than with what we want the new curriculum to accomplish. Although it is important, even necessary, to refine continuously whatever we have in the way of a curriculum, creating a new curriculum by simply refining the old one prevents us from considering distinctly different possibilities and from achieving significant change.

> In looking for ways to improve or create a curriculum, we tend to concentrate more on methods of instruction—how we like to teach—than on the purposes of that instruction. Means customarily take precedence over ends.

> We make curriculum decisions incrementally, grade by grade and subject by subject. The trees concern us more than the forest. Furthermore, the need for a revised or new curriculum to incorporate connectedness over time and across domains of knowledge is honored more in rhetoric than in practice.

> We take the various curriculum domains to be defined by the traditional school subjects and the textbooks that define the substance and organization of those subjects. For example, to judge from the typical school curriculum, natural science would seem to be composed of the disciplines earth science, biology, chemistry, and physics (usually in that order)—and the contents of those separate disciplines are what the textbooks say they are.

> In revising or creating a curriculum, we follow unexamined traditions, rarely

drawing explicitly on research and systematic craft knowledge. The customary experiment-a-week in natural science, for instance, may be the antithesis of real scientific inquiry, yet its place in the curriculum is threatened only by financial duress, not by the lack of empirical evidence demonstrating its value.

We premise our curricula on the day-before-yesterday's technologies, not on today's (never mind tomorrow's). It is as though Boeing set out to create the 777 by assuming propellers and a 10,000-foot ceiling. By thinking of computers and advanced communications technologies as "add-ons" rather than as integral parts of the curriculum infrastructure, the possibilities of major changes in the curriculum are severely limited.

No doubt some improvements have been achieved by these means. But is there a better way? Project 2061 believes that there is. It is to view curriculum as a design challenge and, hence, to approach the creation of curricula more or less in the way in which directors approach the creation of motion pictures and plays; entrepreneurs approach the development of new business; or architects and engineers approach the creation of buildings, vehicles, parks, manufacturing processes, and systems of many kinds.

CHAPTER 1: CURRICULUM DESIGN recapitulates and elaborates on the general design approach described in the Prologue and applies it to curriculum design in particular. CHAPTER 2: CURRICULUM SPECIFICATIONS proposes some important dimensions of a curriculum that have to be taken into account in a curriculum-design effort.

Paul Klee, *Stricken Place*, 1922

CHAPTER 1
CURRICULUM DESIGN

Now let us consider the idea of *curriculum design*. As indicated by the definitions at the beginning of this book, the term "design" is used as a verb to designate a process (as in "designing a curriculum"), or as a noun to denote a particular plan resulting from a design process (as in "a curriculum design"). Never mind that a curriculum is not a garden or a bridge or a traffic pattern; our purpose in this chapter is to see how things play out when we apply the design practices of architects and engineers to the creation of new curricula. And, for the moment, let us put aside the question of precisely what a curriculum is (a matter to be taken up at the beginning of the next chapter), since the *process* of curriculum design can be explored without first having agreement on a precise definition of curriculum.

The purpose of this chapter is to explore ideas, not to provide detailed step-by-step instructions on how to create an actual curriculum design, let alone an actual curriculum. It is as though, by way of analogy, the chapter deals with how general design principles may seem to apply to designing any kind of buildings, but not to how to produce detailed engineering plans for use in constructing actual buildings. To make the argument easy to follow, the chapter parallels the Prologue section by section.

AN INTRODUCTORY EXAMPLE

This time, instead of the backyard garden of the Prologue, our desired end is an effective K-12 curriculum. Our approach need not be altogether orderly, but we would surely do these things:

- We would become gradually clearer on why we want a new curriculum. To have students—some students? all students?—learn more than they learn now? How much more? Learn different things? And which things? To respond to national standards or international comparisons? To raise SAT scores? How far? To increase attendance and the graduation rates? To have more graduates enter college? Certain colleges? To respond to criticism from teachers? Students? Parents? The community? State authorities? All such criticisms, or only some? Which? What if some of our intentions conflict with others?

- And what would get in the way of creating a new curriculum? Teacher, student, parental, or community resistance? Tradition? State laws? Lack of funds? Absence of good evidence for the benefits of change?

- As our goals and constraints become clear, we would identify some alternative design concepts to focus our thinking on curriculum possibilities. A design concept for a curriculum could be to organize instruction around inquiry at every grade level and in every subject, or focus strongly on community issues, or integrate the sciences and humanities, or emphasize the development of lifelong learning skills. On the chance of finding design concepts that may not have occurred to us in the beginning, we would study the curriculum literature, search the Internet, talk to school and university educators who have been involved in curriculum design, and look at curricula in other districts.

- We would narrow the possibilities down to a few appealing design ideas that would work within the design constraints we face, think over our desired goals, and choose an approach that would seem to be the best bet.

- Then we would develop that approach in enough detail to get started actually planning the curriculum. During this stage, trade-offs would have to be considered—a choice, for instance, between the desire to have students cover a large amount of material and to have them develop a deep and lasting understanding of what they study. Our design would describe the structure of the new curriculum, its content, and how it would be operated. In developing the final design, we would hope to call on expertise in curriculum design (books, journals, software, and consultants).

- As actual implementation of the curriculum design progresses, we would come up against unexpected difficulties, forcing us to modify the original design—or to choose an alternate design altogether.

- Even with the curriculum in place, the design challenge would not be over. Maybe the actual curriculum would not match the design very well because mistakes were made in implementing the design. Or the curriculum would match the design quite closely, but we would not get the results we expected. In other words, we would discover or decide that modifications are needed, an eventuality we had anticipated and planned for.

This brief sketch obviously oversimplifies the process of curriculum design even as an ideal, and of course it does not pass muster as a description of what actually happens, if for no other reason than that total K-12 curriculum design is rarely undertaken. But perhaps it suffices to make our main point: the general principles of design used in other fields can apply as well to the design of curricula. On that premise, we now proceed to explore curriculum design in somewhat more detail.

ATTRIBUTES OF CURRICULUM DESIGN

If designing curricula is like designing any object, process, or system in important respects, it follows that it has these attributes:

Curriculum design is purposeful. It is not just to "have" a course of study. Its grand purpose is to improve student learning, but it may have other purposes as well. Whether the purposes are in harmony or in conflict, explicit or implied, immediate or long-range, political or technical, curriculum designers do well to be as clear as possible about what the real purposes are, so that they can respond accordingly.

Curriculum design is deliberate. To be effective, curriculum design must be a conscious planning effort. It is not casual, nor is it the sum total of lots of different changes being made in the curriculum over weeks, months, and years. It involves using an explicit process that identifies clearly what will be done, by whom, and when.

Curriculum design is creative. Curriculum design is not a neatly defined procedure that can be pursued in a rigorous series of steps. At every stage of curriculum design there are opportunities for innovative thinking, novel concepts, and invention to be introduced. Good curriculum design is at once systematic and creative—feet-on-the-ground and head-in-the-clouds.

Curriculum design operates on many levels. Design decisions at one level must be compatible with those at the other levels. A middle-school curriculum design that is

A Project 2061 Glossary for Curriculum Design

Curriculum: An actual sequence of instructional blocks operating in a school. The sequence may cover all grades and subjects (a K-12 curriculum) or some grades and subjects (a middle school science curriculum), and be intended for all students (a core curriculum) or only some students (a college-preparatory curriculum).

Curriculum *Block:* A major component of instruction—from six weeks to several years in duration—that receives separate recognition on student transcripts. Important features of blocks include prerequisites, alignment with benchmarks, and evidence of instruction credibility.

Curriculum *Concept:* An idea that expresses the character of a curriculum design at a succinct, abstract level. Such concepts—usually only from a few sentences to a few paragraphs in length, and perhaps addressing very few aspects of the design—help to focus the design work.

Curriculum *Design:* A proposed organization of particular instructional blocks over time, with instructions for how to navigate among them. Designs—usually described in a few pages—can be invented de novo, elaborated from a curriculum concept, or distilled from an operating curriculum.

Curriculum *Specifications:* A delineation of the goals and constraints to be taken into account in designing a curriculum.

incompatible with the elementary- and high-school designs will almost certainly result in a defective K-12 curriculum, no matter how good each part is on its own. By the same token, the middle-school curriculum itself cannot be effective as a whole unless the designs of its grades are in harmony.

Curriculum design requires compromises. The challenge is to come up with a curriculum that works well—perfection is not its aim. In developing a design that meets complex specifications, trade-offs inevitably have to be made among benefits, costs, constraints, and risks. No matter how systematic the planning or how inventive the thinking, curriculum designs always end up not being everything that everyone would want.

Curriculum designs can fail. There are many ways in which curriculum designs can fail to operate successfully. A design can fail because one or more of its components fail or because the components do not work well together. Or, the people who have to carry it out may reject the design because they misunderstand it or find it distasteful. In most cases, however, curriculum designs are neither wholly satisfactory nor abject failures. Indeed, a key element in curriculum design is to provide for continuous correction and improvement, both during the design process and afterward.

Curriculum design has stages. Curriculum design is a systematic way of going about planning instruction, even though it does not consist of some inflexible set of steps to be followed in strict order. Curriculum decisions made at one stage are not independent of decisions made at other stages, and so the curriculum-design process tends to be iterative, various stages being returned to for reconsideration and possible modification. But recognizing the different tasks and problems at each stage is important in making the process work. The stages, which are considered in turn in the rest of this chapter, are establishing curriculum-design specifications; conceptualizing a curriculum design; developing a curriculum design; and refining a curriculum design.

ESTABLISHING CURRICULUM-DESIGN SPECIFICATIONS

In the fine arts, some creative work can be purely expressive, whatever the artist feels like doing at the moment. Design, though it can be equally creative, is undertaken in a context of purposes—or goals—and constraints. (Even in the fine arts, paintings, songs, and novels usually are more or less designed, not free expressions.) Indeed, some accounts of

In simple situations, designs can be "optimized" to give the best possible outcome on some single variable. But this is a design luxury. In complex situations, it may not be possible to arrive at any design that does better than marginally satisfy all the specifications. The best possible design may not fit any of the specifications well, but attempt only to distribute advantages and shortcomings equitably among all of them.

the design process begin with the design "problem"—a situation involving something that needs to be accomplished with limited means to accomplish it. Goals are the features of a design situation that we desire. If we are not sure what they are, we may not know whether we will be able to accomplish them or not. Constraints are features of the design situation that we cannot avoid. Some will be physical, some financial, some political or legal. If we try to ignore them, they will sooner or later assert themselves—the bridge will collapse, the client will decline to pay, picket lines will be set up, or the sheriff will arrive. Ignoring constraints in curriculum design may eventually have the result that the teacher union will go on strike, the community will replace the school board, or the state will take over the schools. The success of a curriculum-design process will depend heavily on how clearly its goals are laid out and its constraints are recognized.

For the curriculum design being considered in this book, the basic "problem" is to produce science-literate citizens within the limited time and resources that society is willing to provide for the purpose. Moreover, the scope of possible solutions is taken to be limited to what can be done in formal schooling.

Curriculum Goals

The purposes of elementary and secondary education are many. Schooling is expected to foster healthy, socially responsible behavior among young people on their way to adulthood. A modern school system is expected to prepare students for citizenship, for work, and for coping with everyday life, even as it fosters universal literacy and encourages the development of each student's particular interests and talents, whether academic, artistic, athletic, or any other. Accordingly, a lot is expected of a curriculum. Moreover, schools have design considerations—custodial, medical, safety, economic—that are only marginally related to the curriculum as such.

CURIOUS AVENUE   TOM TOLES

In describing a curriculum, whether existing or proposed, the first requirement is that its purposes—*what it is supposed to achieve*—be made clear. Although schooling in general has many purposes, the curriculum is the school district's main instrument for promoting the learning of specified knowledge, skills, and attitudes. Curriculum goals are thus essentially *learning goals*. Strictly speaking, goals are not part of a curriculum—the goals are the ends, while the curriculum is the means—and it is important that the two not be confused. A common example of confusing means with ends is treating "hands-on" activities as curriculum goals in their own right, rather than as one possible means to achieve well-specified learning goals.

To make headway in curriculum design, however, it is necessary to concentrate intensely on the issue of learning goals, identifying those that are credible and usable. To do this properly requires dealing with difficult questions involving what may be termed *investment* (what does it cost in time and other resources to come up with a coherent set of learning goals?); *rationale* (what is the basis for particular sets of goals?); *specificity* (how detailed do the goals have to be?); and *feasibility* (what will students be able to learn?). Wrestling with these questions is worth the time it takes because it will help everyone involved focus on fundamental issues at the very beginning of the effort and maybe even save time in the long run. This section elaborates briefly on the general consideration of the goals discussed in the Prologue, emphasizing some of the particular issues involved in identifying learning goals.

Investment. The process of getting from broad generalizations to grade-level specifics is enormously difficult and time-consuming—at least if it is to be carried out well. It took three years, the direct participation of hundreds of scientists and educators, and multiple levels of review by still other scientists and educators to produce *Science for All Americans*. It took another four years and even more individuals and institutions to transform those adult literacy goals into the grade-level learning goals presented in *Benchmarks for Science Literacy*. The National Academy of Sciences, which was able to draw on *Benchmarks*, took over three years to produce *National Science Education Standards* (which includes other recommendations as well as learning goals). Many states have also invested months and years in creating curriculum frameworks, often basing their work on the national-level formulations of specific learning goals (though with varying degrees of precision). A decade of experience has shown that the meticulous specification of valid learning goals is far different from and vastly more difficult than merely creating one more list of topics to be studied.

These observations are not meant to discourage school districts from specifying what they want a new curriculum to accomplish. Trying to design or redesign a curriculum without clarifying one's goals is folly, for it leaves a district without a clear basis for making design decisions. The familiarity with goals that comes from clarifying each one of them is a significant advantage when the time comes to choose

FROM TYLER TO *BENCHMARKS*

The standard advice on curriculum development was formulated in *Basic Principles of Curriculum and Instruction,* a short 1950 book by University of Chicago professor Ralph W. Tyler. With engaging logic, Tyler asked four fundamental questions:

What educational **purposes** should the school seek to attain?
What educational **experiences** are likely to attain these purposes?
How can these educational experiences be effectively **organized**?
How can we **assess** whether these purposes are being attained?

The Project 2061 plans for redesigning the curriculum are fairly close to this classical formulation. According to Tyler, **purposes** should be derived from the needs and interests of *learners,* features of *contemporary life* (outside the school), and what *subject disciplines* have to offer (to students outside of specialties). This overly large set of possible purposes so derived would then be screened by *philosophy of education* and *psychology of learning.* Philosophy would settle questions such as what values are essential to a satisfying and effective life, whether there should be a different education for "different classes of society," and whether efforts should be aimed at the general education of the citizen or at specific vocational preparation. Psychology would settle questions about whether something could be learned at all, at what age it might best be learned, how long it might take, what multiple purposes might be served by the same learning experiences, and how emphasizing relationships among purposes might lend greater coherence to learning.

Although some curriculum theorists since Tyler have doubted that goals are a good place to begin (or even that they are helpful), their objections seem to be based largely on the difficulty of the task.

The Project 2061 goal specifications in *Science for All Americans* and *Benchmarks for Science Literacy* allow curriculum planners to move directly to the even more formidable task of determining how to achieve these goals. Nonetheless, readers are advised to study what the benchmarks specifically say and what implications they have for materials, instruction, and assessment. Mechanical use of unstudied goals, however good they are, will be unlikely to produce good curriculum. *Designs for Science Literacy* focuses chiefly on the organization of the curriculum, assuming that appropriate educational purposes, experiences, and assessment are in place.

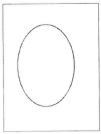

Step 1

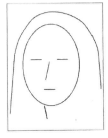

Step 2

Step 3

Project 2061 focuses on a basic core of knowledge and skill for all students. The project is convinced that the basic core so constructed will provide the best foundation for more students to study science even further.

curriculum materials or assessments. And local adaptation of particular goals is more successful when their intent is clear. Moreover, the sense of ownership that develops from the effort to define goals may have important motivational benefits in the hard work that will follow. Clarification, however, does not require starting from scratch.

The popular precept that "all stakeholders should have a hand in setting goals" sometimes is interpreted to mean that goals should actually be formulated locally. But in truth, most school districts simply lack the time and financial resources to do a credible job of creating goals on their own, whereas national groups—and to a lesser degree, state groups—have both. Limited local resources are better employed in modifying already credible sets of goals than in trying to do the work all over again. Moreover, given the mobility of today's U.S. population, it is desirable that local education meets at least basic standards that prepare youth for success anywhere, a scope not ensured by an intense focus on local concerns.

School-district curriculum designers should therefore draw heavily on the work done by national and state groups, and even consider adopting such recommendations in their entirety. The design team should study the recommendations of those groups carefully, making sure they understand the recommendations and the premises underlying them. Then the team can decide whether to adopt them as they are, adopt them with modifications, or do the job themselves. But they should keep in mind that the credibility of a set of goals rests in some large measure on the perceived competence of those who formulated them and on the care that went into their formulation.

Rationale. Whether goals are created locally or drawn from external sources, their credibility depends partly on the rationale offered for the entire set of goals. For example, the rationale used in arriving at the learning goals recommended in *Science for All Americans* was that meeting those goals would benefit graduates by:

- Improving their long-term employment prospects, along with the quality of the nation's workforce, and providing a base for some students to go on to specialize in science, mathematics, or technology or in related fields.
- Assisting them in making personal, social, and political decisions.
- Acquainting them with ideas that are so significant in the history of ideas or so pervasive in our culture as to be necessary for understanding that history and culture.
- Enabling them to ponder the enduring questions of human existence, such as life and death, perception and reality, individual good versus the collective welfare, certainty and doubt.

- Enhancing the experiences of their student years, a time in life that is important in its own right.

Of course, that is not the only possible rationale for the selection of learning goals. Often, only economic or civic purposes are emphasized. And sometimes educators include such purposes as helping students to score well on crucial examinations, secure employment, or qualify for admission to college; fostering general study habits that have lifelong value; or producing graduates with the knowledge and skills that educated adults have had in the past. Some goals may focus not on long-term ends but on short-term means such as lowering the dropout rate or improving the image of the community. Whatever there is to be said for each of these, the only point here is that the rationale for curriculum learning goals ought to consist of a statement of purposes to be served.

Establishing a clear rationale fosters a more thoughtful process of goal selection than arguing each proposed goal *ad hoc*. First, it requires a discussion of how, in principle, goals will be decided. Second, it limits the kinds of arguments that can be made in behalf of a particular goal. Even so, there is not a strict deductive logic linking a proposed goal to one or more rationale statements. Rather, requiring that justification be referenced to an explicit rationale promotes healthful debate by requiring curriculum designers to defend a claim for adopting a particular goal by completing "Everyone should learn this because..." using certain kinds of arguments and not others. A familiar argument that would not pass muster according to the Project 2061 rationale is "Because that is what I had when I was in school and I loved (or hated) it."

Specificity. Expressing curriculum goals in terms of what is to be learned turns out not to be as simple as one might expect. Leaving aside the matter of how to go about deciding on curriculum goals, there is the question of what kind of language to use in characterizing the knowledge and skills that are intended to be acquired, and there is also the question of how specific to be in stating those goals. The greater the grade span of the curriculum, the more difficult it becomes to answer these questions, since the language and specificity appropriate to one level may not be suitable for another. Learning goals can be expressed at many different levels, ranging from very general propositions to very specific ones. The Project 2061 experience in specifying goals provides an example:

The desire for *science literacy for all citizens* led to the general goal that *all students* should be well educated in science, mathematics, and technology by the time they leave their common schooling. This in turn led to agreement on *five criteria* for identifying specific learning goals in science, mathematics, and technology:

utility, social responsibility, intrinsic value of knowledge, philosophical value, and childhood enrichment. Based on these criteria, *Science for All Americans* recommends *65 major learning goals* to be reached by all students by the time they graduate from high school. Finally, *several hundred specifications*, listed in *Benchmarks for Science Literacy*, describe what students should know and be able to do in science, mathematics, and technology by the end of grades 2, 5, 8, and 12.

How far down the specificity ladder to go in framing goals depends in part on how easily a working consensus can be reached among all of those responsible for setting them. Usually, agreement is relatively easy on very general goals and becomes harder as the goals become more specific, if for no reason other than their greater number and their greater demand on technical knowledge. The process should start as general as is necessary and should persistently work toward consensus on sets of specific goals.

Examples of how learning goals have been formulated are to be found in the various national education-standards reports published in recent years. The only explicit discussion of the *language of curriculum goals* in any of them, however, is to be found in *Benchmarks for Science Literacy*. (See sections Characterizing Knowledge and Grain Size from CHAPTER 14: ISSUES AND LANGUAGE of that publication.) The two-page diagram that follows shows a hierarchy of science literacy goals selected from *Science for All Americans and Benchmarks*.

Apart from questions of what level of specificity of learning goals is most useful for purposes of curriculum design, there are questions of what format to use to specify learning goals. A common approach is simply to list headings for topic areas—as general as "chemistry" or as specific as "pH," with little clue as to what would actually be studied and learned under them. A more helpful approach is to express what is to be learned in statements of the knowledge and skills to be acquired by students—*Benchmarks for Science Literacy* and *National Science Education Standards* (*NSES*) being examples in science education. Another helpful approach is to express what is to be learned as descriptions of observable behaviors expected of students—the approach taken, for example, in the *National Standards for Arts Education*. Both statements and behavior formats have strengths and weaknesses that curriculum designers should become familiar with in deciding how best to articulate the learning goals that will be the focus of their work.

A still more detailed approach is to prescribe the exact assessment tasks and criteria for judging them (in a narrow sense, the exact examination questions and scoring scheme). An obvious weakness in adopting such an approach is that those specific tasks alone might determine the curriculum, neglecting students' capacity to apply their knowledge and skill in new contexts.

"Benchmarks was faced with a more difficult language problem in trying to convey accurately what children in the lower grades should learn. It would not do just to match the children's language exactly—Benchmarks is for educators, not students—yet to use the adult technical language of SFAA could encourage teaching it prematurely to children. The solution was to try to say in plain English what the quality of the learning should be and use technical terms only when it was time to make them part of a student's permanent vocabulary."
—*Benchmarks for Science Literacy, p. 312*

See the Bibliography at the end of this book for a list of all of the K-12 standards reports.

CURRENT ARGUMENTS ABOUT GOALS FOR LEARNING

Most everyone concerned with curriculum believes that students learn too little science. (Some educators would claim that students know even less than we think they do.) One approach to solving the problem is to set expectations for student learning higher and higher, in the hope that they will inspire or coerce teachers and students toward higher achievement. Often, this high-expectations approach not only applies to eventual achievement, but also involves pushing expectations to lower and lower grade levels. (For example, third graders may be assigned to study atoms, which is three years before the age when, according to extensive research on learning, children are first able to understand anything important about atoms.)

A different response to lack of student learning is to reduce the shallowness and confusion of an already unlearnably overstuffed curriculum, to make time for better learning of the most important facts, principles, and applications. To the higher-expectations proponents, this approach is "watering down" or "dumbing down" the curriculum. To the better-understanding advocates, the higher-expectations advocates are "elitists" who care mostly about preparing future scientists rather than making sure that all students achieve basic science literacy.

Although often overshadowed by partisan philosophical convictions, the debate requires some underlying facts. What are students currently learning in science? What could they learn under the best conditions? To what extent do expectations that are over students' heads motivate them to learn more than they would otherwise? To what extent will unreachably high demands breed confusion, withdrawal, and learning less than before?

Better knowledge about these issues would help to locate the best trade-off between quantity and quality, to maximize student motivation and minimize confusion. It would be helpful if advocates of both approaches could cooperate on seeking empirical answers to these questions.

The learning goals in *Science for All Americans* and *Benchmarks for Science Literacy* share the same organizational structure, beginning with the three broad domains of science, mathematics, and technology. Within these domains, learning goals are distributed among 12 chapters, 65 sections within chapters, 250 topics within sections, and finally among more than 800 detailed benchmark statements within topics.

ORGANIZATION OF *SCIENCE FOR ALL AMERICANS* AND *BENCHMARKS FOR SCIENCE LITERACY*

Within Domains of Science, Mathematics, and Technology:

Chapter 1: The Nature of Science
The Scientific World View
Scientific Inquiry
The Scientific Enterprise

Chapter 2: The Nature of Mathematics
Patterns and Relationships
Mathematics, Science, and Technology
Mathematical Inquiry

Chapter 3: The Nature of Technology
Technology and Science
Design and Systems
Issues in Technology

Chapter 4: The Physical Setting
The Universe
The Earth
Processes That Shape the Earth
Structure of Matter
Energy Transformations
Motion
Forces of Nature

Chapter 5: The Living Environment
Diversity of Life
Heredity
Cells
Interdependence of Life
Flow of Matter and Energy
Evolution of Life

Chapter 6: The Human Organism
Human Identity
Human Development
Basic Functions
Learning
Physical Health
Mental Health

Chapter 7: Human Society
Cultural Effects on Behavior
Group Behavior
Social Change
Social Trade-Offs
Political and Economic Systems
Social Conflict
Global Interdependence

Chapter 8: The Designed World
Agriculture
Materials and Manufacturing
Energy Sources and Use
Communication
Information Processing
Health Technology

Chapter 9: The Mathematical World
Numbers
Symbolic Relationships
Shapes
Uncertainty
Reasoning

Chapter 10: Historical Perspectives
Displacing the Earth from the Center of the Universe
Uniting the Heavens and Earth
Relating Matter & Energy and Time & Space
Extending Time
Moving the Continents
Understanding Fire
Splitting the Atom
Explaining the Diversity of Life
Discovering Germs
Harnessing Power

Chapter 11: Common Themes
Systems
Models
Constancy and Change
Scale

Chapter 12: Habits of Mind
Values and Attitudes
Computation and Estimation
Manipulation and Observation
Communication Skills
Critical-Response Skills

**TOPIC/SEQUENCE
WITHIN A SECTION:**

BENCHMARKS:

sources of uncertainty

probability

estimating probability from data
   or theory

counts versus proportions

plots and alternative averages

importance of variation and
   around average

comparisons of proportions

correlation versus causation

learning about a whole from
   a part

common sources of bias

importance of sample size

K-2
Often a person can find out
about a group of things by
studying just a few of them.

3-5
A small part of something may
be special in some way and not
give an accurate picture of the
whole. How much a portion of
something can help to estimate
what the whole is like depends
on how the portion is chosen.

6-8
The larger a well-chosen sample
is, the more accurately it is likely
to represent the whole. But
there are many ways of choosing
a sample that can make it
unrepresentative of the whole.

9-12
For a well-chosen sample, the
size of the sample is much more
important than the size of the
population. To avoid intentional
or unintentional bias, samples
are usually selected by some
random system.

Still another approach is to conceive goals as what or how students should study—such as using certain books or works of art, or by employing a "discovery" method of instruction—rather than what students should end up knowing and being able to do. But whatever merits those propositions may have for instruction, they fall outside *Designs'* notion of goals for learning—that is, for what students will eventually know and be able to do.

Feasibility. No matter what resource, rationale, or format is used to solicit suggestions for curriculum goals, it is unlikely that all the goals suggested should be adopted. Almost certainly, there will be too many goals for students to achieve, especially if the main purpose is to design a basic core to be achieved by all students. Priorities must be set by considering what is feasible in the time available for teaching. There is little to be gained, and much to be lost, by expecting more of students than they can possibly learn. A few may be stretched to greater learning, but more will likely just give up or learn to complete assignments mechanically without understanding (or even expecting to understand). But the feasibility line can hardly be set just at levels known to be safely low, for expecting too little will inevitably result in too little learning.

On what basis can learning goals be identified that make sense developmentally as well as conceptually? Teachers and cognitive researchers are the two main sources of pertinent knowledge, as discussed in *Benchmarks for Science Literacy*, Chapter 15: THE RESEARCH BASE:

> *The presence of a topic at a grade level in current textbooks or curriculum guides is not reliable evidence that it can be learned meaningfully at that grade. For example, atoms and molecules sometimes appear in a 4th-grade science reader. Yet extensive research on how children learn about these ideas suggests postponement until at least 6th grade and perhaps until 8th grade for most students. [p. 327]*

> *The single most important source of knowledge on student learning comes from thoughtful teachers. They have firsthand experience in helping students acquire science, mathematics, and technology knowledge and skills. Their input is limited, however, by the realities of the usual teaching situation. Teachers have little time to conduct careful assessments of student learning, lack instruments for assessing richly connected learning and higher-order thinking skills, and rarely have opportunities to compare their experiences with others who teach the same concepts and skills. [p. 327]*

> *Researchers have the advantage of being able to work out a careful design, having time and other resources (including special training in research methods) that teachers seldom have, and undergoing systematic peer review….But research, too, has its limitations. [p. 328].*

Chapter 14 of *Benchmarks for Science Literacy* describes some of the issues related to language and "grain size" used to specify learning goals.

THE FAR SIDE   By GARY LARSON

"Say... Look what THEY'RE doing."

Evidence on learning from both teacher experience and research ought to be interpreted cautiously because it necessarily refers to today's students taught in today's schools by today's teachers. There are so many variables operating in the learning process—teacher and parent expectations, the learning environment, the methods and materials used, the previous knowledge and experience of individual learners, and more—that the failure of students to learn something currently leaves open the question of whether they could have done so if they had had ideal learning conditions from the beginning. [p. 329]

This account suggests that goal-setting groups should draw on the experience of teachers and the findings of research for guidance without expecting definitive answers. It also makes clear why defining goals is not something that can be done casually or quickly, and why it makes sense to depend on the work of those who have the time and resources to bring experienced teachers and knowledgeable researchers together with content specialists to examine the possibilities carefully.

Curriculum Constraints

The other side of setting curriculum-design specifications is identifying the constraints placed on what the design can be like. There are always constraints on design. They may take the form of what will not be permitted and what conditions must be taken into account. As with goals, constraints need to be made explicit if they are to influence curriculum design.

A major impediment to the attainment of curriculum goals is the lack of sufficient time for instruction—hours per day, days per year. In the history of modern education, curricula have been expected to serve more and more goals with few ever being eliminated. An honest acknowledgment of this limitation leads to a conflict among goals: Some goals need to be given up if others are to be met. There is a very real danger that when curriculum committees come up with a delineation of K-12 learning goals, whether their own or adopted, they will fail to purge previous goals that on examination they might find to have lower priority—acting, in other words, as though the time constraint were not real.

Along with time, public discomfort with certain topics (notably human reproduction and evolution) can be a barrier that must be dealt with in curriculum design. Some others are state policies, state and federal legislation, court orders, cost, faculty unpreparedness, lack of suitable instructional materials, standardized tests inadequately aligned to learning goals, college admission requirements, union contracts, and longstanding traditions. And above all, there is the limitation of not knowing enough about student learning—what students can and cannot learn under various circumstances.

What students should know about design constraints at each grade range:

K-2

People may not be able to actually make or do everything that they can design.

3-5

There is no perfect design. Designs that are best in one respect (safety or ease of use, for example) may be inferior in other ways (cost or appearance). Usually some features must be sacrificed to get others.

6-8

Design usually requires taking constraints into account. Some constraints, such as gravity or the properties of the materials to be used, are unavoidable. Other constraints, including economic, political, social, ethical, and aesthetic ones, limit choices.

9-12

In designing a device or process, thought should be given to how it will be manufactured, operated, maintained, replaced, and disposed of and who will sell, operate, and take care of it. The costs associated with these functions may introduce yet more constraints on the design.

—*Benchmarks for Science Literacy,*
pp. 49-52

A rush to accommodate perceived barriers by lowering one's sights should be resisted, however. After all, except for established physical laws, constraints are not necessarily forever. Laws can be changed, budgets raised, traditions recast over time. Curriculum designers should neither ignore constraints nor assume they are insurmountable, but they should try to identify them carefully.

As constraint issues come up at any stage in the design process, it is important to find out what latitude there may be for dealing with them. For example, university admission practices can be looked upon as an impediment to changing the college-preparatory component of the curriculum, for to make major changes may place at risk high-school graduates seeking university admission. But many universities are willing to consider modifying their admission criteria to accommodate school districts, or at least to collaborate with them in systematically trying out proposed changes and introducing them over time if all goes well. Indeed, claimed barriers to curriculum change may turn out to be more an excuse for inaction than a reality. Identifying constraints specifically—as in "We want X but are constrained by Y"—can help to focus a design argument, as in "Do we give up on X or try to eliminate Y?"

Another mechanism for dealing with constraints is to build into the design a process for ameliorating them after the designed curriculum has been implemented. If, for instance, a school district lacks the technological capabilities called for by a proposed design, then instead of abandoning the design entirely, it may make sense to include a plan for technological modernization as part of the design, with an understanding that the new curriculum will be implemented in stages as the school district's technological capacity grows.

In short, as design proceeds, some adjustment of learning goals may become necessary to accommodate constraints that cannot be gotten around, just as some constraints can be modified. Clarifying the specifications for both goals and constraints throughout the entire design process raises the likelihood that the ends and means of the final design will be in accord.

An especially frustrating kind of constraint is one that is locked in multiple ways into a system so no one part can change unless all the other parts change. Such a constraint is sometimes called a QWERTY effect after the first six letters on a standard keyboard, which was designed on a theory of typing that is now outmoded but still almost impossible to change because of the investment in equipment and training based on it. Another example is the English measurement system, which some experts consider to be too woven into U.S. manufacturing to allow us to change to the metric system used almost everywhere else in the world.

CONCEPTUALIZING A CURRICULUM DESIGN

Some overarching idea about the curriculum is a starting place for the creation of a design concept. It may be impressionistic rather than definitive, but no less valuable for that. It provides a point of reference as alternative designs are debated and negotiated. The possibilities are endless, but a curriculum concept commonly includes the instructional contexts to be emphasized, the teaching methods to be used, and the resources to be exploited.

Design concepts can be expressed in a variety of ways—lists and other verbal descriptors, sketches, flow charts and other diagrams, physical models, or accounts of attractive precedents. In the case of a curriculum design, we need at least a brief statement that captures the character of what the new curriculum will be like—or at least articulates separately those few aspects that are deemed to be central.

As we saw in the Prologue, the United Nations governing board informed the architects of the United Nations headquarters in New York City that the new facility should "proclaim the dignity and significance of the infant organization, yet serve as a practical 'workshop for peace,' be international in spirit but still live in harmony with its surroundings, and point to the future rather than honor the past." Such an overarching idea can serve as an inspirational design concept. It provides a point of reference as alternative design possibilities are debated. Similarly, in curriculum design, it makes sense to formulate an overarching idea or a small set of ideas. One way to think about such a statement is to imagine offering a brief answer to this question: What is the curriculum intended to be like?

Below are some examples of possible concepts, not necessarily mutually exclusive, for curricula. Although they promote a variety of goals, any one (or combination) of them would still have to aim also at achieving the agreed-upon set of specific learning goals. These examples of curriculum concepts are not offered as a complete set of categories, but only as a few interesting possibilities that could stand alone or be combined with one another.

- **A classics curriculum** that, in early grades, concentrates on preparing students to study in later grades the great writings, master paintings, musical compositions, grand structures, and scientific discoveries of the ages with increasing understanding and delight.
- **A community-centered curriculum** in which, at every grade level, students explore traditional subjects in relation to community needs and problems, with what constitutes "community" expanding over the years from a neighborhood to a global frame of reference.
- **A high-tech curriculum** that, from the first year on, exploits the power of state-of-the-art information and communications technologies so that all students can become proficient in finding, gathering, organizing, analyzing, and communicating information, which, in effect, would put them in a virtual classroom of worldwide learning.
- **A science and technology applications curriculum** in which all subjects are studied in the context of agriculture, materials and manufacturing, energy sources and use, information processing and communication, health, transportation, and other such general categories of human endeavor.

Architects considered 86 different design concepts for the United Nations Headquarters.

The United Nations Headquarters today.

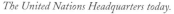

CURRICULUM CONCEPTS

After publication of Science for All Americans *in 1989, teams of teachers, administrators, and curriculum specialists at Project 2061's six School-District Centers began work on unique curriculum concepts that would meet their own local requirements as well as national science literacy goals. They were encouraged to be as imaginative as possible, creating models of what a K-12 curriculum* **could** *look like. Their efforts resulted in some very different curriculum concepts. The teams have characterized their work in the following ways:*

San Antonio Center

This concept is well suited to urban or suburban school districts serving large numbers of students from ethnic and racial minorities. The goal of the concept is to provide all students with school-based experiences organized around the content described in *Science for All Americans* CHAPTER 8: THE DESIGNED WORLD. Students will learn about key aspects of technology in the areas of agriculture, materials and manufacturing, energy, health technology, and communication/information processing. Blocks begin with interesting problems from these content areas and provide at least 16 opportunities for students to participate in designing a solution and/or creating an authentic product. Rather than being tied to specific grades, these opportunities will be designed for four different levels of content complexity and, within those, four different ranges of cognitive abilities.

In pursuing these goals, students will cycle through district learning centers that focus on different technological themes—attending each three times in the elementary-school, twice in the middle-school, and twice in the high-school years. By the junior year of high school, students may choose to continue their progress or begin to work on dual-credit Advanced Placement courses as they begin to shape their careers.

San Diego Center

This concept draws on the city's rich natural and public resources and the economic, cultural, and linguistic diversity of its communities. To provide equitable access to all of the assets available throughout the district, the concept establishes eight regional Resource Centers where students from every community come together to learn. Each Center has a unique theme and focus, drawing on its own regional qualities and resources. Students visit each of the Centers at least once during each grade range.

The curriculum is assembled from a variety of curriculum blocks that have been embedded in the context of one of the Resource Centers. Blocks begin with questions about the world and how it works, and students use evidence to develop and/or evaluate scientific explanations. Through the study of various historical episodes, students learn more about the development of scientific knowledge from tentative hypotheses to rigorously tested theories.

Blocks for each grade range emphasize different, progressive categories of skills and ideas: exploration and discovery for K-2, concept development and research skills for 3-5, relating learning to personal and social issues for 6-8, and expanding perspectives to more global issues in 9-12. Because students spend extended periods of time (from days to weeks) at each of the Resource Centers, it is essential for blocks to integrate or connect with learning goals from the other disciplines. This concept also requires blocks to provide students with opportunities to explore a variety of career options and to meet some career requirements through applied learning experiences.

Philadelphia Center

Designed for a large urban school district with a majority of its students considered to be at the poverty level, this concept organizes learning goals into four major contexts—The Physical Setting, The Living Environment, The Human Organism, and The Designed World—based on *Science for All Americans*. Key characteristics of the concept also serve to describe criteria used to select curriculum blocks. For example, all blocks must reflect the spirit of inquiry, with several blocks at each grade range specifically addressing how scientists work through activities that emphasize problem solving, gathering/analyzing/interpreting data as evidence, controlling for bias, etc.

At each grade range, students explore a different approach to learning: *themes* are emphasized at the elementary level, *issues* at the middle grades, and *case studies* at the high school level. Blocks featuring career-related experiences make up an increasingly greater proportion of the curriculum as students progress from elementary to middle to high school.

San Francisco Center

To respond to the compelling student question, "Why do we have to learn what you are teaching us?" this concept organizes the curriculum around purposeful and contextual learning experiences that are meaningful *from a student's perspective.* All students engage in challenges where they investigate and respond to environmental and social issues, make decisions and solve problems of local and global concern, design and create products and performances, and inquire into "How do we know what we know?" Although challenges can be discipline-based, most challenge-based learning experiences maximize opportunities for students to make connections across disciplines.

To help students learn to think in a "systemic" way, the concept identifies four basic organizing systems—the individual, society, the natural living environment, or the physical universe—that can be used to create challenges that are appropriate to the student's developmental level. Schoolwide challenge-based learning experiences make up only part of the school year, but this pedagogical approach permeates the teaching and learning throughout the educational program.

McFarland, Wisconsin Center

This concept is developed around the assumption that there are at least five behaviors that are inherently meaningful to human beings: stewardship, creativity, wonder, appreciation for the continuum of the human experience, and perseverance. These qualities are woven throughout the 52 projects that comprise the K-12 curriculum.

The curriculum blocks will be cross-disciplinary, open-ended units of instruction called vistas. They will be structured around nine thematic concepts that are based on continuing human concerns: food, water, energy, living organisms, shelter/architecture, exploration, play, recreation, earth/sky, and communication. Each vista is designed to last at least nine weeks and includes multi-age bands of students who work together, learning through inquiry-based activities. Children will stay with the same teams of teachers for each band, allowing a community of learners to develop and grow together. When all 52 vistas are completed, a child will have encountered each benchmark at least twice.

Specific activities in the vistas change depending upon the readiness and abilities of the learners in a particular group. The concept includes a common core of learning for all students, along with time for individuals to follow their own special interests.

Georgia Center

The Georgia concept reflects the unique demands that rural communities (like the three rural counties that comprise the Center) make on schools and their resources. Because of the lack of other local facilities, the rural school is often at the heart of the community's cultural, social, and political activities. Rural students also bring different kinds of knowledge and experiences to school that are unlike those of students in an urban or suburban setting. For example, many children raised on farms come to school already having experienced a "living lab" at home. Even those not on farms often have more space within which to roam and experience nature firsthand. To reflect these unique characteristics, the Georgia concept is designed to help students learn benchmark ideas through "local" topics, including Raising Animals, Timelines, Weather Station, School Garden, Traffic, Where Do I Live?, Communications, Diversity and Independence, Energy, Environment and Human Presence, Evolution, Forces, Human Society and Me, Matter, Part/Whole, Scale, Waves, and Weather and Atmosphere. Many more such learning sequences will be needed to account for all the learning goals outlined in *Science for All Americans*.

- **A hands-on curriculum** in which instruction is largely organized around individual and group projects that favor active involvement over passive learning—in science, actual investigations over textbook study; in art, studio work over slide lectures on art history; in social studies, preparing reports on actual community problems; and so forth.
- **A language-immersion curriculum** in which a standard liberal-arts curriculum is invested with the development of language competence that facilitates the participation of Americans in global business and in cultural and scientific affairs.
- **A learning-to-learn curriculum** in which, in every subject and at every grade level, learning techniques are emphasized even more than the acquisition of given knowledge, guided independent study is featured as a way to develop these techniques through practice, and graduation is based on the student's showing competence as a self-learner.
- **An individualized curriculum** in which, in the upper grades, each student fashions—from a rich array of diverse offerings—a personal program of studies in collaboration with parents and guidance counselors, and in which graduation is predicated on the student's completing that program and passing examinations in prescribed subjects.
- **A work-study curriculum** in which academic studies are leavened with supervised real-work assignments in school (teaching, cafeteria, gardening, building maintenance, clerical, etc.), or in the community as volunteers (in nursing homes, parks, libraries, university and industrial laboratories, etc.), so that students develop good work skills and a commitment to community service, in addition to receiving a basic education.
- **A "vistas" curriculum** in which instruction is organized into a relatively few, interdisciplinary, cross-grade settings—such as a farming plot, a forest site, or community service operation—in which students participate several times, at different levels of sophistication, over their K-12 school careers.
- **An inquiry curriculum** in which, at every opportunity, study is motivated and organized by students' own questions and efforts to find answers themselves.
- **An environmental curriculum** that uses the description and operation of the physical and biological environment—and the social issues associated with them—as a focus for learning all subjects at every grade level.

In these few examples, each curriculum-design concept features only one or two aspects of a curriculum. Of course, a complete final design has to incorporate all aspects of the curriculum as a system, but the drive to create and promote a new curriculum commonly comes from an inspiring emphasis on just one or two of its dimensions. In any case, after considering several possible curriculum concepts—there are always alternatives—one must be selected for development.

Judging from the language found in the education literature, the need for general characterizations is widely recognized. One trouble with such shorthand designators is that they are often no more than popular slogans of the day and only superficially characterize curricula. It may be hard to distinguish a curriculum claiming the banner of "hands-on," "problem-solving," or "back to basics," from one that does not. Something more than a label is needed.

Still, it is not particularly helpful to have a long treatise on one's philosophy of education, particularly since such statements tend to encompass political and instructional issues as well as curriculum, and often have a tenuous connection to the actual curriculum design. What stands to be most useful as a design concept is no more than a paragraph or two—more than a slogan, less than an essay—setting out the main ideas, themes, or features that help to make sense out of what might otherwise appear to be a hodgepodge, a curriculum without character or personality.

DEVELOPING A CURRICULUM DESIGN

Once progress has been made toward setting curriculum goals, identifying design constraints, and selecting a design concept, the main task of developing a full-fledged design can proceed. Curriculum design calls for making decisions on what the content and structure of a curriculum will be. Because the task is a complicated one, it is well for the school district's curriculum-design team to consider what strategies it can use to facilitate the process.

Curriculum-Design Strategies

Developing a K-12 curriculum design is difficult because a curriculum is complex and because the tools available for creating one are few. Three basic strategies that can help include copying or modifying an existing design, compartmentalizing the design challenge by grade range or subject area to make design development manageable, and testing aspects of the emerging design in the development stage. Any combination of these strategies can be used.

Copying or modifying an existing design. To design a curriculum from scratch is possible, but difficult. Invention is hard and usually not necessary. Instead, it makes sense for a curriculum-design team to look for an existing curriculum design that could be used to meet many, if not all, of the goal and constraint specifications. If the search is successful, the question then becomes whether to copy it as is or to modify some of its features. The problem with implementing this strategy is that although there are plenty of K-12 curricula in operation, there are virtually no adequate descriptions of them that can serve as designs to follow. (Prose about curriculum tends to be rhetorical rather than operational.) Chapter 4: CURRICULUM BLOCKS suggests the possibility of curriculum "models"—more than just design concepts, but much less detailed than complete designs—that would eventually be available to guide local design.

Calvin and Hobbes
by Bill Watterson

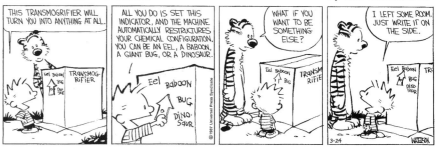

In the absence of adequate descriptions of curricula, an actual, operating K-12 curriculum can be studied to develop such a description. Using this approach, the curriculum-design team identifies another district's curriculum that seems to be something like what it has in mind. The team studies that other district's curriculum and student performance and perhaps visits its schools to interview teachers, administrators, board members, students, parents, and community leaders. If its own district is reasonably similar to the other district being studied, the team may decide to adopt the entire other curriculum or some of its properties. If, as a consequence of a series of such investigations, features are adopted from several different districts, the team will then need to reconcile them with one another. This will be a time-consuming activity but no different in principle from designing a new school building instead of a new curriculum. In fact, however, few districts are likely to undertake such an expensive and problematical effort.

Compartmentalizing the design challenge. A K-12 curriculum is a complex system. But such complexity can be dealt with by focusing on only those aspects of the curriculum that have to be designed afresh and by breaking the task into more manageable parts and dividing the work among different subgroups of the design team. The obvious question is how best to compartmentalize the effort. It may not matter much if later steps are built into the process to ensure that the earlier parts get put back together in a cogent whole. A danger in dividing the design effort by *grade range* is that the resulting curriculum design will not be coordinated well between grade ranges. If the design work is divided by *subject matter*, the resulting design may lack intellectual coherence. Or if the work is divided by categories of students, the design may produce social conflicts. And so on—no one way of dividing the design effort is entirely satisfactory.

One possibility for safeguarding overall curriculum coherence is to have a cascading committee structure for the design team. Suppose, for instance, that two committees are formed in each *school:* (1) a cross-subjects committee that pays attention to how subjects relate in each grade in that school, and (2) a cross-grades committee that pays attention to how each subject is developed from one grade to the next. And then there would be *districtwide* cross-subjects and cross-grades committees with representation from the school committees. And, above that, there would be a single central committee made of representatives of the districtwide committees. And finally, representatives of *that* committee would serve on a board-appointed, community-wide curriculum-design advisory committee.

Such a formidable committee structure may sound oppressive in the telling, but it does have the virtue of involving many teachers in the process and of forcing the need for coherence to the forefront. However, it would be a great help to be able to begin the effort with a set of specific learning goals already arranged for conceptual coherence over K-12, and to have available curriculum blocks with known internal coherence (as discussed in Chapter 4: CURRICULUM BLOCKS). It should also prove helpful to have computer software that could keep track of the cumulative characteristics of curriculum blocks as they are progressively assembled into an entire curriculum.

Testing in the development stage. No matter how many committees are involved, how thorough the analysis that takes place, and how persuasive the arguments that are made, things can go wrong. Good design practice calls for testing elements of an emerging design even as design development is under way. Curriculum features can be tested in single schools, in single grades, in single classrooms, or even with subgroups of students within classrooms. And combinations of these tests are possible, such as subgroups of students in single classrooms in each of several schools. Not every aspect of a curriculum can be tested on a limited scale, since much of a curriculum depends upon interactions among its component parts. But some ideas and propositions do

Early experiments in transportation

lend themselves to testing, and a search of the literature sometimes reveals that some have already been put to a more or less rigorous test.

Curriculum-Design Decisions

In all design work, several concepts are employed in making decisions. As outlined in the Prologue, these are benefits and costs, risk, and trade-offs. Here they are examined from the perspective of curriculum design.

Benefits and costs. When goals are settled on early in the design process, the benefits expected from a new curriculum remain fairly stable. Costs do not. Every component of a K-12 curriculum has costs associated with it. Teachers, instructional materials, facilities, and support staff cost money; instruction takes time. Costs, however, are not unambiguously tied to benefits. Some curriculum components cost more than others. Inquiry-based science, for instance, costs more than textbook-based science, but how much more? What learning accrues from the former that is missing in the latter and vice versa? Is the benefit—added learning—worth the added cost? And to whom?

Decisions about spending on facilities and resources often derive more from speculation and fashion than from good evidence of effect on student learning. Money is a severely limited resource in most school districts, but an even more critical cost factor is time. Like money, time invested in one component of a curriculum is time not available for other components. However, there is a nearly absolute ceiling on time available for teaching and learning: currently about 1,000 hours a year for 13 years. Student time is finite; so is teacher time and so is time in the school class schedule. Hence, the argument for inclusion of any proposed component must be that its learning benefits justify the time it will take. Could more important things be learned in the same time? Can the same learning be gained in less time? Time and dollar costs provide estimates of effectiveness when weighed in relation to expected learning outcomes.

Political costs cannot be avoided, since most important curriculum decisions are viewed differently by different individuals and groups. Unanimity is hard to come by in education. Teaching methods, grouping practices, course requirements, promotion and graduation requirements, etc., almost always attract both support and opposition. In meeting the design challenge realistically, it is not possible to avoid all social costs, but it is essential to avoid having all of the curriculum decisions favor certain individuals and groups over others.

Risk. The notion of risk may be obvious enough in the case of industrial products, but in education we are not used to thinking in terms of risk except, perhaps, the risk that something will not have the benefits claimed for it. In fact, though, any curriculum design does have risk associated with it. For one thing, there may be undesirable, even unanticipated, side effects. For instance, introducing abstract topics early in the curriculum may increase the risk of many children failing to learn the material, losing self-confidence, and staying away from the subject in later years. Also, small-group instruction may help some students with certain learning styles to improve their understanding but impede other students. And courses that require a lot of homework may penalize students who come from disadvantaged home situations.

Another risk commonly connected with curriculum change is whether the change will pay off with higher scores on standard examinations—or will instead lower scores by attending to goals different from those reflected on the tests. And there is the "transcript risk": four years of, say, "integrated science" in high school may not pass muster in some college admission offices without the requisite labels "chemistry," "physics," and "advanced biology."

Risk cannot be avoided. Good design practice calls for making an effort to ascertain what risk is associated with various curriculum propositions. This approach reduces—but does not eliminate—the chance that unanticipated, unwanted side effects will occur, and it alerts curriculum-design teams to look for ways to reduce the identified risks. Consistent with the notions of "overdesign" and "redundancy" mentioned in the Prologue, for example, instructional systems can provide safety nets for students who do not succeed within the allotted time and instructional resources. At the very least, risk analysis alerts educators and parents that particular dangers are associated with the proposed curriculum.

Trade-offs. Curriculum could be designed more confidently if benefits, costs, and risk could all be quantified. For example, if all benefits could be expressed in a common measure (say, the average number of new ideas learned per student) and all costs could be estimated in terms of a common measure (say, dollars or curriculum-hours), then it would be possible in principle to maximize the ratio of benefits to costs and risks. But there are no such common measures. How can the benefit of better student self-esteem be compared to the benefit of better multiplication? What dollar cost can be assigned to teachers having more time for teaching and less time for their personal lives? Still, at the least, it is useful to list benefits and costs and estimate gross differences in priorities among them as a basis for making plausible trade-offs.

The trade-off concept acknowledges that there are no perfect solutions. There are only solutions that, compared to one another, bring different benefits, different costs, different risks. Thus A and B may appear to deliver about the same benefits, but B costs less and A is safer—so the decision will hinge on making a trade-off between low cost and low risk. Or, C may result in more learning on average than D, but be less effective for students in need of special help and for those with unusual talent. There the trade-off is between student populations, and whichever way the decision goes, the designers are on notice that some action will be needed to compensate the group placed at relative disadvantage.

Benefit-cost-risk analysis is not widely used in making curriculum decisions, at least not explicitly so. Most propositions put forth to modify existing curricula try to make a case for the absolute value of the proposed change: this or that course is needed to accomplish this or that purpose, or increasing the time allotted to this or that subject is inherently good. Sometimes a comparative case is made: it is better to introduce such and such in the 4th grade than the 5th, or all students now taking general mathematics should take algebra.

Such propositions are not sufficiently tough-minded. Effective trade-offs can be made only when questions such as the following have been answered: What learning benefits that would be missed otherwise will accrue to students? Which students? What evidence is there for that claim? What will those benefits cost? Are the benefits worth the cost? Can the same benefits, or nearly the same benefits, be acquired more cheaply? What risks are there associated with the proposed action? Who might gain and who might lose? These kinds of general questions, and others that pertain to local circumstances, should at least be entertained. Though there is no adequate calculus for balancing them, thinking about them can at least reduce unpleasant surprises—and may reveal unexpected opportunities.

REFINING A DESIGNED CURRICULUM

A curriculum rarely works as well as its design would lead us to expect, and some need for tuning is inevitable. With luck, some of the needs for tuning will be identified during the final stages of design. Inevitably, some will show up only when the design is fully implemented. And, beyond components not working quite as planned, every design is likely to have unexpected side effects. Even if the curriculum works well enough for awhile, eventually some things are likely to go wrong and need fixing. And in time, no matter how smoothly the curriculum is functioning, its design will become obsolete. New knowledge, new methods, new technologies, and new circumstances will open up new possibilities. So a design should include provisions for monitoring the implementation of the curriculum and its effects.

Aspects of a curriculum design for which systematic monitoring is desirable emerge from the premise that (1) the actual curriculum matches the design, (2) the students subject to the curriculum are actually acquiring the learning the curriculum has been designed to effect, and (3) the learning, once acquired by students, is having the benefits attributed to it. It is also necessary, of course, to monitor whether the various costs are within tolerable bounds—but that is an aspect of change that schools are already able and eager to perform.

Curriculum Congruence with Design

To make judgments about a curriculum *design*, a school district needs to know the degree to which the *actual* curriculum is a reasonable rendition of that design. We cannot make valid judgments about a given aircraft design if the manufacturer deviated from the design in significant ways—and such deviations are much more likely to occur in the "manufacturing" of a curriculum than an aircraft. This suggests that periodically, particularly in the early years, a new or revised curriculum be checked for its match to the intended design. This can be done in two complementary steps, internal and external.

Internal. Committees composed of teachers, administrators, students, and interested citizens, including some members of the design team itself, should be established to monitor assigned aspects of the implementation process. For example, data can be collected on the time allotted to instructional blocks, on the patterns of enrollment in them, and on the comprehensiveness of specific learning goals ostensibly targeted by them. A cross-grades, cross-subjects oversight group can study the committee's findings and prepare a report for the board of education. Such a study should be made annually until there is confidence that the curriculum matches the design—either because the implementation has been faithful or because the design has been modified to match practice. Afterward, internal studies should be conducted on a specified schedule, say every three or four years.

External. All institutions need input from external perspectives. School systems are complex institutions whose parts, including the curriculum, ought to undergo periodic examination by outside experts. The tradition of "visiting committees," common in college and secondary education, is increasingly common among grade schools as a way to obtain impartial but authoritative opinions on how well a curriculum matches the adopted curriculum design. This policy, if budgeted for on something like a four-year cycle, is well within the means of most school districts, and it acknowledges that all technological systems require feedback and control to operate as intended.

Both of these methods for checking the congruence between an implemented curriculum and its design depend on having an explicit description of the curriculum design. Such a description makes it possible for reviewers to know what to pay attention to and helps them avoid going off on low-priority tangents.

Learning Results

Assuming the curriculum has been correctly implemented and is being well operated, questions remain: Are students learning what the curriculum design intended they would? Are they learning some things but not others? Are all categories of students learning what is intended, or only some of them doing that? Is adequate learning occurring at every grade level? In every classroom?

To answer such fundamental questions requires detailed prescriptions for what is to be learned. In the domain of science, mathematics, and technology education, *Benchmarks for Science Literacy* provides a basis for estimating student learning for specific levels—at the 2nd-, 5th-, and 8th-grade levels. *National Science Education Standards* can also be used for that purpose in science, as can the National Council of Teachers of Mathematics' *Curriculum and Evaluation Standards for School Mathematics* in mathematics, and similar standards for technology education.

Whatever benchmarks or learning standards are used, the point is for agreement to have been reached, *before* a new curriculum is instituted, on how to measure learning and where the checkpoints will be. Learning measures should be derived from—or at least be demonstrated to match—the learning goals set at the beginning of the design process, and the checkpoints should be keyed to the grade-range decisions made in the early part of the design effort. Since the purpose is to estimate the effectiveness of the curriculum design, not judge individual students, the evaluation can be spread over time. A three- or four-year cycle, examining different subject domains in different years, and sound sampling of both goals and students, will reduce the investment of time and money necessary to conduct the studies.

The results of such studies constitute performance profiles to hold up to the preselected benchmarks so that decisions can be made about modifying the curriculum design. Where discrepancies are found, the question will arise as to whether the curriculum design is at fault in some general way or whether one of its components is internally inadequate for its designated role. For example, inadequate learning of concepts about biological systems detected at the 8th-grade checkpoint after a substantial biology course in 7th grade may mean that there should not be any such course that early, or only that the particular instruction materials or teacher preparation for that course weren't good enough. The purpose of periodic studies is to raise just such design questions.

Long-Term Consequences

Meeting specific learning goals is the first test of a curriculum, but it is only a near-term measure. For most things, curricula no less than bridges and buildings, it is the long term that counts. Does the learning serve the learners well? To some degree that can be determined from benchmark testing: If students meet the 8th-grade benchmarks, does that appear to put them in a position to do well in high school? But eventually the horizon of interest must extend beyond the school years, for the main purpose of schooling is to prepare young people for an interesting and productive adulthood.

It is not realistically possible for a school district to mount the kind of longitudinal studies that give precise information on the effects of its curriculum. Perhaps, as research on teaching and learning advances, better short-term indicators may be found for long-term effects. But even so, there are simply too many variables at work to isolate a few and hold others constant as the most credible scientific study of effects requires. There are statistical techniques for studying multiple-outcome variables and adjusting for differences in inputs, but sophisticated multivariate studies are very costly.

Networks of schools may be able to pool resources for such studies. It is more practical, however, to carry out opinion surveys of graduates periodically. Survey questions for graduates (as they encounter life, jobs, or further schooling) could cut across the entire spectrum of school experience, including some pertaining to the curriculum. If done sufficiently early after graduation, this feedback should be of use in considering what adjustments in the curriculum may be needed.

"Planning Ahead"

LOOKING AHEAD

In this chapter, the general ideas about design derived in the Prologue have been applied to curriculum design. The propositions necessarily have been general, with little on the actual process of creating a curriculum design. Yet to be considered are how goals should be adopted, how constraints should be identified and dealt with, how choices should be made among alternative curriculum concepts, what properties of a curriculum need to be considered, what trade-offs should be made, and the like. But recall that the purpose of the chapter is to lay out the idea of curriculum design and not to serve as a blueprint for action. Further, the intent so far has been to provoke discussion among teachers and others on just how to respond to the question: What is involved in designing an entire K-12 curriculum? Chapter 3 considers the dimensions of the curriculum, focusing on those that are most important to the design process.

Rhonda Roland Shearer, *Geometric Proportions in Nature, Study No. 1*, 1987

CHAPTER 2
CURRICULUM SPECIFICATIONS

Although the general principles of design may apply widely, as has been argued, they have to be shaped to respond to the kind of thing to be designed, whether a ballet, a housing development, or any other object, event, or system. How can we best characterize the essential features of the K-12 curriculum? In answering, the chapter starts with a more or less standard definition of curriculum and recasts it in more structural terms. Curriculum structure, content, and operation are discussed, and a case is made along the way for developing and using curriculum graphics to facilitate thinking about those salient aspects of curriculum design.

WHAT IS A CURRICULUM?

Judging by how people talk about it, "curriculum" may be thought of as anything from what is written down in official district documents to what actually goes on in classrooms day to day. To complicate matters, curricula are often spoken of in terms of one or another of their special features (such as liberal arts, Great Books, language-immersion, activity-based, assessment-based, and—these days—standards-based curricula), in terms of students tracks (giving us college-preparatory, vocational, and "general" curricula), in terms of subject matter (the reading, mathematics, and Spanish curricula, for instance), and much else.

In books on K-12 "curriculum," a curriculum is usually treated as a collection of courses, where a "course" is an educational unit usually at the high-school or middle-school level, consisting of a series of instruction periods (such as lectures, discussions, and laboratory sessions) dealing with a particular subject. "Courses" are usually a year or a semester long, but quarter or trimester courses and courses spanning several years are becoming more common. In earlier grades, curriculum is more typically described

curriculum

1. the whole body of courses offered by an educational institution or one of its branches.

2. any particular body of courses set for various majors.

3. all planned school activities including courses of study, organized play, athletics, dramatics, clubs, and home-room program.

— *Webster's Third New International Dictionary*

Since all of us have extensive experience as students in school, we all have a strong sense of what makes up a school curriculum...academic subjects, which are cut off from practical everyday knowledge, taught in relative isolation from one another, stratified by ability, sequenced by age, grounded in textbooks, and delivered in a teacher-centered classroom....This shared cultural understanding of the school curriculum exerts a profoundly conservative influence, by blocking program innovations even if they enhance learning and by providing legitimacy for programs that fit the traditional model even if they deter learning.

— D. F. Labaree, "The Chronic Failure of Curriculum Reform" (1999)

in terms of "subjects." Courses and subjects are themselves often subdivided into "units," which run from only a few days to a few weeks. For purposes of designing an entire K-12 curriculum, the component parts should be quite large—more like a course in extent than like a teaching unit. This idea is described further in CHAPTER 3: DESIGN BY ASSEMBLY and CHAPTER 4: CURRICULUM BLOCKS.

Dictionary definitions of curriculum are usually compatible with the "scope" part of "scope and sequence," a phrase commonly used in education to mean that a description of a curriculum must say what the curriculum is made up of and how it is arranged. More specifically, a curriculum is always an assembly of instructional components distributed over time, never a single component in isolation. A chemistry course, for instance, is not a curriculum, though it may be an element of, say, a high-school college-preparatory curriculum or of a university premedical curriculum. And the collection of components forming a curriculum is a configured set of studies, not a haphazard collection—some things come before other things, some serve one purpose, others another, some are intended for all students, others for only some students, and so on.

Since a curriculum always has boundaries, a description of one should make clear what educational territory it encompasses. The boundaries of a curriculum can be identified according to grade range (an undergraduate curriculum, a K-12 curriculum, a middle-school curriculum), content domain (science, music, language arts), and student population (a core curriculum, a vocational curriculum, a college-preparatory curriculum, a bilingual curriculum, a prelaw curriculum). A "Project 2061 curriculum," could be said to be any assembly of K-12 science, mathematics, and technology instructional components designed to enable all students to achieve science literacy as defined in *Science for All Americans*.

The design of curriculum within its prescribed boundaries involves a variety of aspects that we consider in this chapter under the headings of structure, content, and operation. Obviously there is likely to be interaction among the categories—decisions about structure have to take some account of the demands of content and the limitations of operation and vice versa—though such interactions are not dealt with explicitly here.

CURRICULUM STRUCTURE

Whereas architecture deals with the configuration of space, curriculum deals with the configuration of time. Minutes and years are for curriculum design what inches and miles are for architectural design. In a sense, there are two time dimensions to a curriculum: clock time and calendar time. Clock time is the number of minutes allotted to instruction each day, typically portioned into distinct periods for different subjects or courses. Calendar time is the duration in weeks and months (or quarters or semesters). Together, these two temporal dimensions of school determine a total amount of instruction time that can be considered a "curriculum space" to be filled.

To start, it is extremely important to settle on the overall dimensions of a curriculum, since it is often useful to partition the whole into components for planning purposes. Whether the K-12 curriculum should be designed as a whole or divided into parts depends on the stage of design. Consider three possibilities shown in the diagrams below. In each, the horizontal dimension represents the 13-year calendar span of the school curriculum and the vertical dimension implies the daily instruction time available.

The bottom diagram here provides the comprehensive K-12 perspective needed to achieve a totally coherent curriculum, but is too large for most practical planning. Yet,

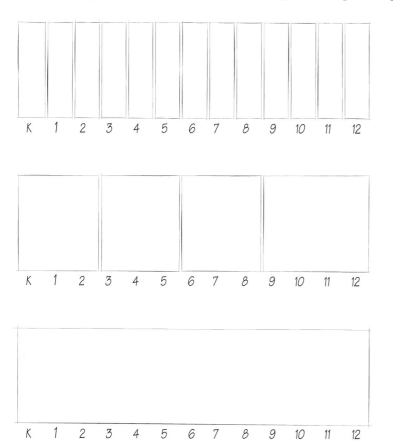

The Project 2061 publication *Atlas of Science Literacy* is an important tool for relating parts of a curriculum to the whole. It maps how student understanding of key ideas would grow and make connections over the entire K-12 span.

curriculum designers must have a way to refer back from the parts they are working on to the whole—which is why having a clearly stated K-12 curriculum concept is important. At the other extreme, the top diagram (on the previous page) indicates that planning can be done separately grade by grade. The trouble with this, of course, is that it is almost certain to lead to a fragmented K-12 curriculum. A reasonable middle ground is to plan within several grade ranges, as depicted in the center diagram. The center configuration has several advantages for planning:

- A span of three or four years seems manageable.
- The sections can be assembled to give a picture of the entire K-12 span.
- The properties of single grades can be inferred from the three- or four-grade spans.
- The four ranges approximate developmental stages—early childhood, late childhood, early adolescence, and adolescence (or the early elementary, upper elementary, middle-, and high-school grades).

Still another reason for using such curriculum spans is that benchmarks and content standards are arrayed in that way. While *Benchmarks for Science Literacy* uses the four divisions displayed above, *National Science Education Standards* (and also the standards in some other school subjects) use three ranges—K-4, 5-8, and 9-12. Although *Designs for Science Literacy* uses the four-part set of boundaries, the ideas presented are equally applicable to planning in three (or five) grade ranges.

Once agreement has been reached on the grade ranges for planning, one can turn to general structural features within them that are fundamental in some sense, but do not yet deal with subject-matter content. The parallel in our garden example in the Prologue was in deciding how much of the garden area to reserve for flowers, vegetables, and trees without getting into the details of which particular flowers, vegetables, and trees.

What then can be considered structural in curriculum design? We propose three main structural properties—without any claim that there are not other possibilities—that raise basic questions about the nature of the curriculum being designed: What balance is sought across the curriculum between core studies and elective ones? How much variation in time blocks is acceptable? What instructional formats should be included?

Core and Electives

A key structural feature of a curriculum is the distribution between core studies and electives. "Elective," it is important to note, does not necessarily mean only studies that go beyond basic literacy. Core components of a curriculum are those in which all students participate, not necessarily the venue in which they achieve all of the common learning goals. A well-designed set of electives could provide for different students to achieve some of the same learning goals in different contexts. The basic literacy goals concerning force and motion, for example, might be achieved by some students in a transportation elective, by other students in an environmental elective, and by still other students in an astronomy elective.

The distribution of core and elective curriculum components can be represented graphically. But it may be difficult in practice to decide what in the curriculum is actually core and what is not core. It is clear enough what is core when all students must take the same course at the same pace and with the same requirements for success; it is a little less clear when all students must take the same course, say 10th-grade biology, but are grouped in such a way that different students have different versions of it; it is less clear still in cases in which students are required only to take the same subject, say "mathematics," but under that title may have very different courses—perhaps algebra or business math. But such differences are a part of what is involved in structural analysis—analyzing how common the core really is will likely promote serious discussion of some fundamental issues.

There are curricula in which all students take exactly the same program of studies (not altogether rare in private schools and the lower grades of public school systems); and there are curricula in which each student chooses a unique path, with no common core (more difficult to find in practice). But the great majority of curricula have various proportions of studies that are core and noncore electives.

Suppose that after a committee charged with designing a curriculum has reached a working consensus on the general character of the curriculum being designed, subcommittees for each of the four grade ranges work out plans. Let us say they decide as follows:

- Grade range K-2 will be all core, meaning all students have the same program of studies.
- Grade range 3-5 will be all core, but there will be options for students to pursue the topics at a more advanced level, though at pretty much the same time and location—call it "core-plus."
- Grade range 6-8 will also be core-plus but in addition it will reserve about 20 percent of the span for alternative electives scheduled separately, gradually increasing them each year.
- Grade range 9-12 will reserve about half of the first two years for core-plus (with more in 9th grade) and after that all electives except for a single capstone course required of all seniors.

In different situations, a group of people designing curriculum might be called a design team, committee, study group, or still other combinations of such terms. In this book those terms are used more or less interchangeably, as seems convenient. In different situations also, curriculum design might be undertaken by a set of schools within or between districts, rather than by a school district per se.

Graphically, the results of the subcommittee deliberations may be rendered as

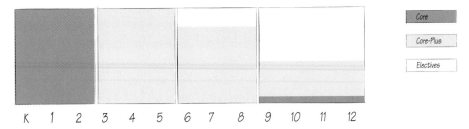

A grade-by-grade arrangement to achieve such proportions might look like this:

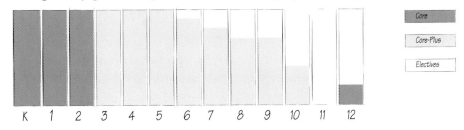

Schedule Variation

As things stand now in most curricula, the components are nearly uniform in time structure: they are either a semester or a year long, and all periods in the day have the same number of minutes. Presumably the period of approximately 45 minutes was settled on as a good compromise among various considerations—including the attention span of students and the number of subjects that have to share the time. The considerable advantage of uniform divisions of school time is that they can be neatly and reliably coordinated—their beginning and ending times are synchronized and are the same every day, so students can easily enroll in a variety of different subject combinations. Transitions from one to the next can also be uniform and minimal, simplifying the task of keeping track of where students are. Students can mix and match subjects.

Some educators argue, however, that not all content fits a given time container equally well. The U.S. Department of Education report *Prisoners of Time*, which claims that the greatest barrier to curriculum reform is the misuse of time, particularly faults the uniformity and rigidity of instructional time blocks. The report notes that, especially in the upper grades, all school subjects are almost always either a year or a semester long, meet every day of the week, and are allotted the same number of minutes per meeting.

Prisoners of Time: Report of the National Education Commission on Time and Learning, 1994.

In any case, in current practice it is seldom the intrinsic demands of a subject that determine the size and shape of its time dimensions, but the other way around—the time dimensions are set and each subject must do the best it can to fit the time available. Think what it would be like if all containers in a grocery store were required to be the same size and shape, and every product—bread, eggs, milk, watermelons—had to be made to fit them, no matter what. The containers would stack nicely, but would require a considerably awkward fit for some contents.

Laboratory experiments and design projects are prominent activities that require set-up (and take-down) time and so would obviously benefit from longer—and fewer—divisions of time. In "block scheduling," some periods are given double length or more, allowing greater flexibility within each period. In an extreme version, a single period could fill a whole school day—or even a week. Such blocking raises significant issues in deployment of staff. If extended periods were devoted to single subjects, teachers would have to plan for longer (but intermittent) activities. If extended periods were shared between different subjects, as in various brands of integration, staff would have to plan more cooperatively. The structural question here, then, is how much variation will be permitted in the time subdivisions of a curriculum.

This is not to argue for either uniform or variable curriculum configurations, nor is it to suggest that curriculum designers should make the spaces first and then fill them up with content. Rather, it is to focus attention on the need in curriculum design to decide on what time constraints will have to be met—how much variation will be permitted.

This may, of course, differ by grade level. Consider a case in which grade-range design subcommittees end up proposing the following:

- Grade ranges K-2 and 3-5 both decide that individual teachers will be permitted to devote different amounts of time to different subjects within prescribed limits. If, for example, 20 minutes per day of science were required on average, which would amount to 60 hours in a school year, that time could be scheduled for 20 minutes every day, or an hour twice a week, two hours once a week, a half-day every other week, or, in an extreme example, all day for two solid weeks once a year.

- Grade range 6-8 opts for uniformity, with all classes meeting for one period every day for a semester or a year, thus keeping everyone's schedule simple and facilitating changing classrooms for special subjects.

- Grade range 9-12 calls for all courses to be a semester or a year long, but different subjects can meet for three or five periods a week (or, in the case of laboratory, studio, and shop courses, for seven periods a week). The number of periods

Additional discussions of schedules appear in:
Chapter 3 in the Candidate Blocks and Configuring Blocks sections
Chapter 4 in the Time Frame subsection
Chapter 6 in the Alternative Time Patterns section

a day for each course can also vary. However, an integrative core course required of all students is required to meet one period each day for the entire four years.

This arrangement may be represented graphically as

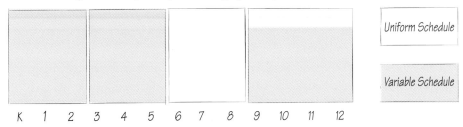

Uniform Schedule

Variable Schedule

K   1   2   3   4   5   6   7   8   9   10   11   12

In Chapter 3, more detailed attention is given to a variation in actual partition of school time into courses and units.

Instructional Format

A K-12 curriculum as traditionally viewed is composed of "subjects" in the lower grades and "courses" in the upper grades. Instruction strategy changes somewhat with grade level, but by and large it is built on cycles of homework, recitation and class discussion, lectures (sometimes disguised as class discussion), hands-on activities (such as demonstrations, laboratory experiments, short projects, and field trips), occasional independent study and seminars, and eventually quizzes and tests.

For some kinds of learning tasks and in some circumstances, the traditional organization of instruction can be effective, especially when there is a well-defined body of easily understood content. Traditional instruction in the form of subjects and courses has a long history and, as materials and technology have been improved over the years, such instruction has arguably become more successful—at least in the hands of properly prepared and supported teachers. As teachers become aware of how superficially students can learn some ideas and of how persistent students' misconceptions can be, and as they become more adept at applying cognitive principles of teaching and learning to their instruction, the "traditional" lecture-discussion or lecture-discussion-laboratory formats may be used more effectively for benefiting a wide range of students. Innovations within traditional formats, such as cooperative groups, self-paced study, and computer-based instruction, have offered additional possibilities for increased instructional effectiveness. Even so, there is reason to ditional format is satisfactory for many kinds of learning, no mat-

r formats exist that may be better for certain purposes, such as for
ability to participate effectively in group discussions or for learning on
ple, in addition to traditional courses in typical daily time frames, a
le stand-alone *seminars* that meet once or twice a week, stand-alone
occupies all of a month or longer, and stand-alone *projects* that
or years—not as minor parts of courses, but as free-standing major
12 curriculum that have their own entries in student transcripts.
these three formats. Educational research has not produced consis-
of them materially improves student learning, but they still appear
likely to play roles in instructional development in the future.

small number of ideas from multiple perspectives, there is
g a seminar format. A Socratic seminar is not simply a dis-
people, but a method for sharing in the examination of read-
k. Seminar texts include news or magazine articles, novels,
speeches, research reports, or epic poems and can be in sci-
ne, literature, politics, philosophy, or any other field.
ne who understands the format and has some background
n the exact subject itself may not be required, especially
bject-matter experts to intervene in the process to make
rrectly understand the content, rather than to guide the
ome of the best seminar leaders may not yet be familiar
ide a model for how to ask questions and consider alter-
ning in leading seminars, many people other than teach-
ng them are parents and other community members,
, and even students at higher grade levels who can be
ding seminars for students in lower grades.
ir time demands. Generally, they should meet only
rticipants time to study the source materials and pre-
hey can last anywhere from a few weeks to a semester.
a regular course led by the regular teacher, but a semi-
may be reduced by the role the teacher is tempted to

SHARING GOALS WITH STUDENTS

Presumably all instruction has specified goals, but the goals may not be clearly shared with the students. For example, in a project to design a catapult, the students' purpose is to build a winning catapult, whereas the teacher's intent is for them to learn about constraints and trade-offs in design.

Tests can give students a pragmatic indication of what the goals at least *were*. To the extent that students are advised on the nature of the assessment tasks—and how they will be scored—they may know the goals early enough to work deliberately toward them. (Hence the ubiquitous student query, "Will that be on the test?") But assessments themselves are not always well suited to the underlying goals of instruction.

Independent study. As adults, we are pretty much on our own to learn what we need or want to know when we want to know it. We may do so by taking courses, reading, listening to lectures or tapes of lectures, using computer programs, asking experts, and so forth. In all these cases, we guide our own instruction. It is puzzling, therefore, that much of K-12 schooling (and undergraduate education, for that matter) provides little opportunity for students to learn and practice how to be independent learners. Students in school are told by teachers just what to study on a daily basis, what pages to read from a specified textbook (that in turn signals them in boldface type and glossaries which words to memorize), what experiments or other activities to carry out, and which end-of-chapter problems to do.

Development of independent learning skills can be fostered by explicit curriculum provisions for doing so, such as courses that include major independent-study components, and stand-alone blocks of time for independent study that are not part of a traditional course. One form of independent study is *goal-specified assignments*. Students are told what knowledge or skills they are expected to learn, what resources are available to them, and what the deadline is for accomplishing the learning. (See the nearby box about sharing goals with students more generally.)

Students should not, however, be sent off entirely on their own. Research has long shown that, for independent study to work, students need monitoring and coaching on how to proceed and on what is to be learned. Successful completion of the independent study assignment requires that the student submit a report or takes an examination (written, oral, or performance). As with seminars, independent study assignments are sometimes embedded in traditional subject and course formats. A possible advantage of having stand-alone independent study may be that teachers who are particularly good at coaching such study could specialize in it.

Projects. A distinctive form of independent study is a project to be carried out by a single student or by a small team of students. Projects can be of any duration (though a deadline should probably always be set), and can have an inquiry or action orientation. A project can stand alone as a curriculum entity or may be part of a course, but in either case it should overseen by a person acting as a coach rather than as a traditional teacher.

A particularly promising kind of project provides an opportunity for students to teach each other. Before being permitted to carry out the task, the would-be peer teachers should demonstrate their own competence in the material to be taught and must have their teaching plan approved by the project adviser and by the teacher of the target students.

In short, courses typically follow a textbook and are operated by teachers; seminars are based on readings and other non-textbook materials and are operated by seminar leaders who need not be regular teachers; and independent study takes the form of goal-specified assignments or projects, which are overseen by project advisers and coaches.

Thus, another structural property of the curriculum that can be depicted graphically is the proportion of instruction to be allotted to different formats. Suppose, after arguing the merits and drawbacks of these instructional formats in the light of the overall learning goals set for the new curriculum, our grade-range subcommittees decided the following:

- Grade range K-2 will be entirely traditional, with separate periods of time for mathematics, science, etc.
- Grade range 3-5 will introduce the equivalent of about three periods a week to independent study; the rest of the curriculum will be traditional in format.
- Grade range 6-8 will be composed of about two-thirds traditional instruction; the rest will be independent study, including peer-teaching projects.
- Grade range 9-12 will divide the curriculum into 50 percent traditional, 30 percent independent study, and 20 percent seminars.

In graphic form:

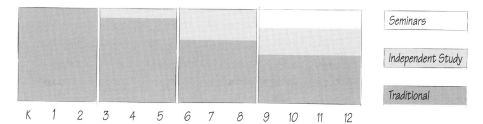

These proportions could, of course, be realized by a variety of grade patterns. For example, the high-school format could be met, as these diagrams suggest, by treating each year the same, by changing the proportions each year, or by concentrating the seminars and independent study in the last two years:

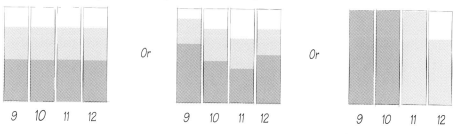

CURRICULUM CONTENT

In a neat design process, agreement would be reached on the general features of a curriculum before considering how content will be organized within those limits. In most practical situations, however, some back and forth between general features and particular organization is highly likely. Just as in our garden example, where we may have to decide which particular trees and shrubs to plant in the area reserved for them and how they would be placed, so curriculum designers have to decide upon the composition and arrangement of subject matter over time. Decisions also have to be made with regard to which principles for organizing subject matter are preferred—at one extreme organizing content by disciplines, at the other extreme by completely integrated studies, or by some mix of discipline-based and more-or-less integrated studies.

Content Distribution

To curriculum designers, the large-scale layout of subjects is of more interest than the details of what topics are to be treated in what fashion. Such a layout is analogous to beginning the design of a hospital by indicating roughly where the various facilities, wards, private rooms, emergency rooms, laboratories, and business offices are to be located, without specifying precisely how many of each there will be or how they will be equipped. A garden designer could begin by describing the general location of perennials, annuals, shrubs, fruit trees, and vegetables without indicating exactly which particular varieties there will be in each location. As more detail is added to the design, the character and purpose of the hospital or garden become evident.

The content of a curriculum can be dealt with on different levels of specificity. At the most general level, the issue is the relative attention paid to the major domains: arts and humanities; science, mathematics, and technology; and other common studies (vocational, physical education, health, and business). Suppose that with regard to those three categories (in that order):

- Grade ranges K-2 and 3-5 decide to divide the curriculum into 50 percent arts and humanities (emphasizing reading); 40 percent science, mathematics, and technology (emphasizing arithmetic); and 10 percent other (health and exercise).
- Grade range 6-8 opts for 40 percent-40 percent-20 percent, thereby increasing the attention given to health and physical education.
- Grade range 9-12 agrees to increase the "other" category to 30 percent by including vocational and other noncore subjects.

This could be represented graphically as

Other	Other	Other	Other
Science, Mathematics, & Technology	Science, Mathematics, & Technology	Science, Mathematics, & Technology	Science, Mathematics, & Technology
Arts & Humanities	Arts & Humanities	Arts & Humanities	Arts & Humanities

K   1   2   3   4   5   6   7   8   9   10   11   12

From such a broad demarcation, the content distribution can be decided in progressively greater detail. Each of the three major domains given above can be examined in further levels of specification. Examples of successive levels of detail for the preceding example of 9-12 curriculum are portrayed below.

Going from left to right, the first level shows how a 9-12 curriculum may configure time generally; the second how the science, mathematics, and technology portion may divide time among science, mathematics, and technology; and the third shows how the science part may allocate time among broad science domains. These diagrams indicate how time will be apportioned among various content categories but not how the content will be organized conceptually or in what sequence it will appear.

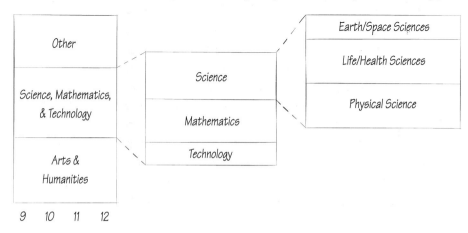

Content Organization

From the smallest teaching unit to a multiple-year sequence of courses, content is expected to be more than a jumble of topics. Lesson plans, course outlines, and curricula are each expected to be made up of content that conceptually forms a coherent whole. The coherence of an entire curriculum requires, of course, that the parts of which it is built have their own internal coherence. But even if each of the components of a curriculum is *internally* coherent, the curriculum as a whole may not be. In other words, curriculum coherence means that, at any level of content organization, the parts have to make sense in view of the whole and vice versa.

Although several different styles of coherence are possible, traditionally coherence is assumed to be provided by the internal organization of the separate disciplines or fields, usually as they appear in the respective introductory textbooks used in college survey courses and imitated in high-school courses. Disciplines, however, are not fixed. They evolve, although not smoothly or in an altogether predictable direction, and they occasionally undergo radical change. They overlap and intermingle—and, sometimes, new disciplines emerge. But for purposes of design, it is sufficient to think of a *discipline-based curriculum* as one organized on the basis of the knowledge, methods, structure, and language of one or more of the academic disciplines.

But recently (although not for the first time), some educators have urged turning away from basing the K-12 curriculum on the individual disciplines. They claim that, whatever their value for research, the disciplines are too compartmentalized, abstract, and remote from the interests and concerns of most people living in a complicated world to serve the general education needs of students. It would be better, they argue, to *integrate* parts or even all of the curriculum across fields and disciplines, organizing the curriculum around interesting phenomena, important cross-cutting themes, design

PEANUTS CHARLES M. SCHULZ

projects, or urgent social and environmental issues. Content integration can take place at a high level of generality (science and art, for instance), within a broad domain (such as science, mathematics, and technology, or any two of those), or between areas within disciplines (algebra and geometry, or physics and biology), but what distinguishes an integrated curriculum is that something other than the disciplines determines how the content will be organized.

That is not to say that discipline-based curricula necessarily neglect environmental issues, say, or that integrated curricula disregard knowledge and methods from the disciplines. The same set of specific learning goals could be pursued within either form of organization. The difference is more of a foreground/background situation. In a discipline-based curriculum, particular academic disciplines or fields catch our eye first (with applications of one kind or another coming in to view from time to time), whereas in an integrated curriculum, phenomena, themes, or issues are out front (with disciplines behind the scenes). Cogent arguments have been made for both approaches (as evident in the articles cited in the Bibliography), but there is little empirical evidence for any advantage in results of one over the other.

Still, design calls for decisions to be made. Should the curriculum be discipline-based, integrated, or partly discipline-based and partly integrated? In making such decisions, both content and pedagogical issues have to be taken into account, and clearly specified learning goals and constraints are essential. How will a strictly discipline-based curriculum ensure that students reach the thematic, historical, and other nondiscipline-based goals? How will an integrated curriculum ensure that they reach the learning goals in the physical sciences, life sciences, earth sciences, mathematics, and technology?

Taking all of this into account, it is useful to consider what would happen if a school district were to decide to do the following:

- Design a curriculum in which there is substantial but not complete commitment to content integration, and the integration will be mostly at the science/mathematics/technology level rather than the science/arts/humanities level.
- Integrate K-2 science, mathematics, and technology around phenomena of interest to very young children, rather than having a separate period of time for each subject.
- Treat mathematics separately in grades 3-5, as well as integrating it with science and technology around themes such as "scale" and "change."
- Separate the disciplines in grades 6-8, dividing the available time equally among them.

- Integrate science, mathematics, and technology in grades 9-12 around social and environmental issues for a third of the time and let students select one or two disciplines to pursue in depth for the rest.

Graphically, these decisions for the core curriculum in the domain of science, mathematics, and technology can be represented in this way:

K 1 2	3 4 5	6 7 8	9 10 11 12
Science, Mathematics, & Technology	Mathematics	Algebra/ Geometry	Biology or Chemistry or Physics or Engineering or Astronomy/ Geology or Statistics
	Science, Mathematics, Technology	Physical Science	
		Earth Science	
		Life Science	Science, Mathematics, & Technology
		Engineering	

In a discipline-based organization, disciplines can appear one after the other or in parallel—that is, more or less concurrently. In series, students study each subject in turn for a considerable period of time, usually every day for a semester or year. In parallel organizations, students study all of the target subjects more or less simultaneously. (It would be rare, however, to study, say, physical *and* biological sciences every day—"concurrently" usually means more rapidly alternating from one day or week to the next.) Although this arrangement allows connections to be made among the disciplines, it still keeps them front and center. In a thematically integrated curriculum— say, one that focuses on lakes or spacecraft design—the disciplines may become indistinguishable and so sequence becomes truly parallel. Note that some curricula, despite their titles, are not actually integrated. A common example is middle-school general science, which often turns out to be a rotation of the individual science disciplines on a semester or six-weeks basis—essentially a series sequence.

The distinction between curriculum sequences can be portrayed as these hypothetical patterns of science courses in a single range:

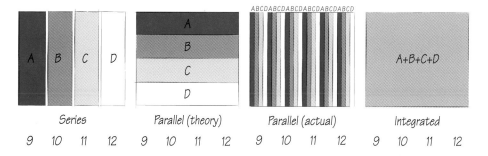

Series	Parallel (theory)	Parallel (actual)	Integrated
9 10 11 12	9 10 11 12	9 10 11 12	9 10 11 12

The typical high-school science curriculum is configured in series. So is the arrangement proposed by a group of scientists and science teachers in Chicago, although the traditional sequence is reversed. In the *Scope, Sequence, and Coordination* type of curriculum proposed by the National Science Teachers Association there would be a parallel configuration in which students study four natural sciences every year for four years, the organization of topics within each science coordinated with the others in mutually supportive ways. In the following diagrams depicting these three arrangements, the shaded areas are what essentially all students take, and the unshaded ones are electives:

| Earth Science | Biology | Chemistry | Physics | Traditional |

| Physics | Chemistry | Biology | | Chicago Plan |

| Earth Science
Biology
Chemistry
Physics | Scope, Sequence
& Coordination |

9   10   11   12

CURRICULUM OPERATION

Notions of how the finished products will be operated are built into the design of gardens, bridges, and buildings, and care is called for in anticipating how any such product will actually be used once it is off the drawing board. For example, an aircraft design requires consideration of the number and functions of crew members, how maintenance will be provided, and how passengers will board and exit the craft. These operations effectively become part of the design and are difficult to modify after the aircraft is in production.

So too with curricula. Four key features affect the operation of a curriculum and need to be taken into account in curriculum design: *student pathways* through the curriculum, *staff deployment*, the selection and use of *instructional resources* (including decisions about technologies), and *monitoring* and maintaining the effectiveness of the curriculum. Each is discussed briefly below. Most of the operational issues concern educational philosophy or limitations of resources. They are listed here only as questions that have to be argued and decided in the design process, but discussion of the many issues involved will be left to the considerable literature devoted to them.

Student Pathways

In designing a curriculum, it is necessary to identify the ways in which students will progress through their K-12 years. That information can be developed by posing a series of design-related questions:

See the Bibliography for chapter-by-chapter references to relevant readings. *Designs on Disk* contains a number of these readings for easy access.

- Will all students follow one path? If there will be more than one path, how many will there be? How will students be grouped on any one path? On what basis will students be placed on one path or another? Will they change paths at any point in their passage through the curriculum, or only at certain points?
- How will students advance through the curriculum—grade by grade or grade range by grade range? Will advancement be automatic, or will promotion be based on demonstrated performance?
- What are the criteria for a student to enter or exit a particular curriculum subject or course?
- What is required for graduation?

Staff Deployment

The following set of design-related questions can be used to identify staffing needs and resources:

- At what point in the elementary-school curriculum will teachers be expected to be subject-matter specialists? In which subjects? Will secondary-school teachers need to be specialists in a broad domain, such as science, or a specific discipline, such as chemistry?
- In how many different grades or grade ranges will teachers have to be proficient? Will they have to cycle through the grades, or specialize in one or two?
- What skills other than those of traditional classroom teaching will teachers be expected to have: project coaching, seminar management, supervising independent study, overseeing peer teaching, training and supervising noncertified teachers, or others?
- Will the curriculum design permit teachers to specialize in one or two such functions (in contrast to subject-matter specialization)?
- Would it be legal to have students or uncertified adults conduct some of the teaching called for by the curriculum? Will teachers connected to the school only by television, the Internet, or regular mail have recognized status as faculty?

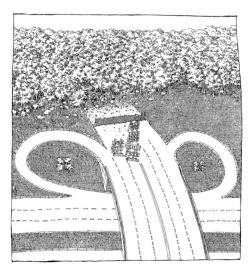

"Small Victory-Highway"

Instructional Resources

The following questions are aimed at identifying the need for and availability of such resources:

- If courses depend heavily on textbooks, how will the books be selected to ensure that they match the learning goals of the curriculum? If curriculum blocks do not use textbooks, how will the needed materials be identified, reviewed for relevance and accuracy, and selected? How will staff be trained to use them effectively?
- Will the curriculum operate with whatever spaces and technologies are available, or will it presume the availability of certain information and communications technologies? If so, which ones?
- If the curriculum will require the use of advanced technologies, what demands will that put on the deployment of teachers and the design of school facilities?

Curriculum Monitoring

The following design-related questions are intended to identify curriculum-monitoring needs and resources:

- How will we know whether the curriculum is having the intended effects? What will be the criteria for student performance? How often will major student assessments be made and at what checkpoints? Who will judge what the findings imply?

What will be done with the results to ensure that deficiencies are corrected?

- What measures will be taken to detect unwanted and unanticipated side effects that may occur between student assessments? If it is known that the design may put some students more at risk than others, what special arrangements will be made to monitor their progress? What will be done about teachers who don't adapt well to the design?
- What provisions will be made to monitor the financial, time, and political costs of implementing the curriculum design? What contingency plans will be in place if the cost of operating the curriculum exceeds estimates by an unacceptable amount?

SUMMING UP

In Part I we have considered the ideas of curriculum design in particular and proposed a way of thinking about curriculum that takes into account key properties that come into play across the entire curriculum. These properties— structure, content, and operation—can be summarized briefly by the questions they raise about a curriculum:

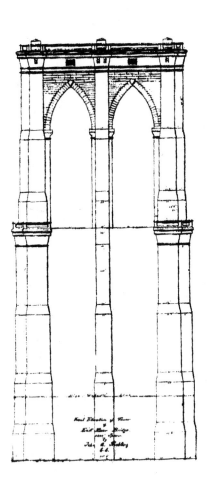

Structure
- What is the distribution between the core studies that all students must take and electives?
- Do all subjects have the same time configuration? If different time frames are permitted, what are they?
- What is the pattern across the curriculum of traditional instructional formats and alternatives such as seminars and independent study?
- Where are curriculum checkpoints?

Content
- What are the specific goals for student learning?
- Is content organized by discipline or is it integrated? If discipline-based, which ones? If integrated, at what level and on what basis?
- Is content arranged in series or in parallel sequences?

Operation
- What pathways through the curriculum are open to students, and how is it determined which students follow which routes?

- What capabilities do the staff need to have, and how are staff to be deployed?
- What resources are essential to operate the proposed curriculum?
- What provisions are built into the curriculum to find out if it is having its intended effects and not having unwanted ones?

Attention should be given to all of these issues from the beginning of the design process. Nevertheless, it is clear that the answers will be shaped in part by smaller-scale decisions that are made along the way, as actual curriculum components are considered and chosen. Not only must individual components—courses, for example— have their share of the desired properties, but collectively they must fit together into a coherent whole that will satisfy the specific goals for learning. The next set of three chapters proposes an approach to the design of a complete curriculum by selecting and sequencing components that have well-specified properties, particularly those outlined in this chapter.

Conceivably, a team of curriculum designers could undertake fixing, gathering, and constructing instructional components—lessons, activities, and units—to fit the specifications of the kind laid out above. In what follows, *Designs* proposes two other possibilities: Part II presents a long-term alternative based on the assumption that resources will eventually become available to make possible the local design of whole curricula; Part III suggests, how, in the short term, smaller-scale but still significant improvements in curricula can be undertaken as part of building capability for the long-term design venture.

PART II
DESIGNING TOMORROW'S CURRICULUM

A basic premise of *Designs for Science Literacy* is that there will come a time early in the 21st century when it will be possible for any school district to design its own curriculum in its entirety, from kindergarten through the 12th grade. A district will be able to create a curriculum to advance the goals it values, incorporate the teaching approaches it believes will work well with its students, and use time and other resources as it wishes. It will be able to choose or construct a traditional or radical curriculum, or one that is conservative in some respects and innovative in others. It will be able to do that and still have an intellectually coherent, developmentally sound curriculum that meets national and state standards.

Currently, it is not possible for most school districts to undertake such sophisticated curriculum designing. The design tools are not available, most of the necessary curriculum-building materials have not been created, and the very notion of designing a whole curriculum—not just improving instruction in parts of the curriculum—is far from being widely accepted. The three chapters that make up Part II assume that these constraints will gradually be removed in the early years of the new century. The chapters also assume that, as with almost everything else, the curriculum design process will be computer-assisted.

A central proposition of these chapters is that significant responsibility for directing the process of K-12 curriculum design will be vested in teams of teachers. In creating a design, teachers will, of course, consult with parents and community leaders, university professors, materials developers, district curriculum specialists, colleagues in professional associations and scientific societies, and even with professional curriculum architects. Though legal authority for the curriculum has traditionally rested with school-district boards and state agencies, teachers have had the main responsibility for its design and implementation, and that arrangement is expected to continue. The mark of a mature profession is that its practitioners have substantial decision-making authority over fundamental aspects of their work. On the other hand, teacher teams will not be held responsible for the extremely demanding task of creating the instructional materials that a curriculum will need.

There is little said about instruction in the three chapters in Part II. Although having good instruction day by day may well be as important as having coherent curricula, it is not necessary in curriculum design to spell out repeatedly what good teaching practices are. Both *Science for All Americans* and *National Science Education Standards* provide principles that ought to characterize K-12 teaching. But here, in the three chapters of Part II, it is assumed that instructional strategies will have been built into the blocks from which curricula will be configured. The chapters themselves focus on the nature, selection, and assembly of such curriculum blocks.

CHAPTER 3: DESIGN BY ASSEMBLY presents a concept of curriculum design by selection and assembly of already developed curriculum blocks, according to design specifications. It also suggests how computers can be exploited in the service of curriculum design and curriculum management. CHAPTER 4: CURRICULUM BLOCKS describes the properties of curriculum blocks, and how a large pool of them could be created. CHAPTER 5: HOW IT COULD BE offers three fictional accounts of how curriculum design could proceed in the early 21st century, assuming that the possibilities presented in Chapters 3 and 4 do indeed become realities.

The three chapters that make up Part II need not be read in order. In particular, some readers may find it helpful to look first at Chapter 5 to get an idea of how the ideas in Chapters 3 and 4 would eventually play out in practice, before returning to the details of blocks and how they may be assembled. The Introduction to Part III includes suggestions on practical approaches to reform that are also relevant to Chapter 5.

John Cedarquist, *Cabinet*, 1989

CHAPTER 3
DESIGN BY ASSEMBLY

One can argue that making a series of small improvements in existing curricula will eventually add up to new curricula that accomplish what is desired of them, including producing graduates who are literate in science. That seems an excessively optimistic view. Decades of making incremental adjustments in K-12 curricula have not resulted in much advance toward universal science literacy.

If incrementalism has not worked, what will? Surely not instant revolution, since the record of radical curricular reform is no more impressive than that of gradual curricular evolution. The central issue is not speed, but how to get a broadly beneficial transformation to occur at all, however long it takes. The traditional way we go about curriculum change is simply not up to the job. Present curriculum design fails to focus on the attainment of specific learning goals, is piecemeal, expects teachers to design curriculum materials, pays little attention to validating learning, and is technologically archaic.

Typically, what *is* done in the name of curriculum design is to modify slightly the large elements that are already there—by adding units and topics to existing courses (sometimes subtracting, but not often), changing teaching materials and teaching methods, and altering the rules governing the paths that students take through the curriculum. Thought of as adjustments made to improve a curriculum, such changes may make good sense individually, but they still leave unattended the need to create a basic configuration of subjects and courses in the first place.

There must be alternatives to the piecemeal approach. The one presented in this chapter is to design curriculum by selecting and configuring large curriculum components called blocks, almost necessarily with the aid of computer software.

THE IDEA IN BRIEF

Imagine there exists a large and diverse inventory of "curriculum building blocks" and a database describing each of these blocks in detail. Given explicit learning goals and constraints and a conception of what the curriculum *as a whole* should be like, a school district could create its curriculum by choosing and configuring sets of appropriate blocks chosen from the inventory. The developers of the blocks would bear the responsibility of building in sound goals and instruction, and the makers of the database would bear the responsibility of studying and describing blocks well, leaving local educators with the responsibility for making good choices.

A Familiar Analogy

By way of comparison, imagine designing a sound system for your home. In the early days of radio and records, the choices were simple: Depending on your pocketbook and whether you wanted a floor model or tabletop model, you bought a single piece of furniture and plugged it in. If you also wanted to play records, you got a separate phonograph. But then "high fidelity" and "stereo" arrived and with them the notion of buying individual audio components; hook the right components together in the right way and you had a sound system tailored to your personal tastes and circumstances.

The drawback of this technological advance was that there were now more choices to be made, more things to be considered in making those choices, and more ways of going wrong. To deal with this new complexity, some people would simply pick a recommended set of supposedly matched components sold together by a catalog or audio store, letting someone else make most of the detailed decisions. But others, rising to the challenge, would decide to design their own sound system.

Today, stores and catalogs are crowded with audio components—many different makes and models of loudspeakers, amplifiers, tuners, disc players, and tape decks, each with its own specifications and cost. To design a system, you need some rules for making choices. Most of the rules are simple. You must have a least one signal source (radio receiver, disc player, tape deck, etc.), or you may have several. You must have an amplifier and speakers (or headphones) if you prefer. You must pay attention to technical details like the relationship between the power output of the amplifier and the power needs of the speakers. In deciding what components to get, you also need to take into account the kinds of recorded material you expect to play, what physical constraints there are (such as the size of the room and who else lives nearby), how much you are willing to spend, and how you want the system to look (how it looks, after all, is an important source of motivation and satisfaction—given that it produces good sound). For example, some people want a separate preamplifier, power amplifier, and tuner, whereas others prefer a receiver that combines all three functions in one box.

Once you have all the components at home, you must connect them properly. Beyond selection criteria, you need rules regarding connections. You follow instructions, hook up the system, plug it in, turn on the power—and likely as not it doesn't work. You consult the instructions and diagrams, change some connections, poke some controls, and eventually it usually does work. You like the sound at first, but later you realize that although it works, you are not satisfied. You seek advice from

Technology moves fast. Today, video and computer components are often part of a "home entertainment system" which may be more than a "sound system."

audio magazines or experts or friends, upgrade some components, and add video components to create a complete home entertainment system. And it is a very different system indeed from the radios and phonographs of yore.

In principle, there is an even more challenging alternative to using a set of components. With the right technical knowledge and skills, you could create a sound system from transistors, resistors, wire, transformers, and other electric, electronic, and mechanical parts. (A generation ago, some people bought do-it-yourself kits to undertake this time-consuming challenge.) Few of us today have the technical expertise or time for that, especially with the advances that have been made in electronic technology, including the incorporation of a great many small components into integrated-circuit chips. So our realistic choices come down to either obtaining a preassembled sound system or assembling a system from major components.

The design–by-assembly process for developing a curriculum has much in common with designing a sound system. Both enterprises involve selecting components that are already available and putting them together in a particular way. Both require that the decision makers choose from a variety of existing components on the basis of what the completed system will be expected to accomplish, as well as the performance characteristics, reliability, compatibility, and cost of the individual components. And both make success contingent upon the decision makers' abilities to make good choices.

An Education Analogy

To focus more particularly on education systems, think of a university and ask yourself what its undergraduate curriculum is. Your answer may well be that it has many curricula because each undergraduate designs an individualized curriculum from a huge collection of courses. The university catalog tells what courses are available and specifies the selection rules for particular programs.

Selection rules provide first-year students with essential information, such as: (1) each course carries a certain number of credit units; (2) to graduate, students must accumulate a certain number of credit units; (3) these units must be distributed over time within some predetermined constraints; (4) students must select a major and meet its course requirements; (5) students must pay attention to course sequence, for some courses cannot be taken until prerequisites have been satisfied; and (6) students must also pay attention to weekly and daily class schedules, so that their courses don't overlap. (For examples, see the excerpts from a Harvard University course catalog shown opposite.)

In choosing courses, savvy undergraduates do not depend altogether on the selection rules given in the catalog. They seek out student opinion on who are the best teachers, which are the tough courses, and the like. Gathering relevant information, official and otherwise, is part of the design process. In short, students design their individual undergraduate curricula by selecting instruction blocks (courses, seminars, independent study) from a defined set according to some explicit selection rules, taking into account what is known about the available courses. Even then, most students find through experience that they have to modify their initial choices. They usually do so by changing single courses or course sections, but they may make more radical changes, such as choosing a new major.

This analogy is not intended to suggest that elementary- and secondary-school students should design their own curricula, but to show that the process of curriculum design based on assembling the right components in the right order is not, after all, a novel idea.

TYPICAL DESCRIPTIONS FROM A HARVARD UNIVERSITY COURSE CATALOG

Earth and Planetary Sciences 6
Introduction to Environmental Science: The Solid Earth
Catalog number: 2694
Primarily for Undergraduates
Half course, Fall, Tuesday and Thursday, 10-11:30; lab and section require one afternoon per week.

An introduction to geology, with primary emphasis on those aspects of continental near-surface phenomena whose understanding is particularly relevant to environmental problems and hazards. Environmental effects of natural subsurface processes (plate tectonics, earthquakes, volcanoes) and surface processes (erosion, deposition, mass movements); resource use (water, soil, minerals, fossil and alternative fuels); waste disposal. Labs and field trips familiarize students with minerals and rocks, geological structures, and maps, and the interpretation of field observations.
Note: EPS 6 may not be counted for a degree in addition to EPS 7.

Engineering Sciences 50
Digital Electronics in Scientific Experimentation
Catalog number: 4499
Primarily for Undergraduates
Half course, Fall, Tuesday and Thursday, 10-11:30.

Intended to give students in laboratory sciences and students contemplating a concentration in electronics a thorough grounding in the concepts and language of digital electronics as well as some experience applying these concepts in practice. Topics include analysis and design of combinational logic circuits, sequential logic circuits, state machines, programmable logic devices, and the essentials of analog signal conditioning techniques. "Hands-on" experience in the use of integrated circuits is provided by a combination of experiments done with a take-home lab kit, and some exercises using laboratory equipment and computers. A miniproject is assigned during the reading period.
Note: Some experience in a laboratory science is helpful but not required.
Enrollment: Limited to 36.

Mathematics Xa
Introduction to Functions and Calculus: A Year-long Course I
Catalog number: 1981
Primarily for Undergraduates
Half course, Fall, Section I: Monday, Wednesday, and Friday, 10:00; Section II: Monday, Wednesday, and Friday, 11:00; Section III: Monday, Wednesday, and Friday, 12:00; twice-weekly lab session to be arranged.

Fundamental ideas of calculus are introduced early and used to provide a framework for the study of mathematical modeling involving algebraic, exponential, and logarithmic functions. Thorough understanding of differential calculus promoted by yearlong reinforcement. Applications to biology and economics emphasized according to the interests of our students.
Note: Students taking Mathematics Xa should plan to take Mathematics Xb immediately afterwards. (The sequence Xa, Xb is equivalent to the Mathematics Ar, 1a sequence.)
Enrollment: Limited to 15 students per section.

Anthropology 97x
Sophomore Tutorial in Archaeology
Catalog number: 0400
Primarily for Undergraduates
Half course, Spring, hours to be arranged.

The sophomore tutorial provides a background in archaeological method and theory, particularly focusing on small-scale societies. Specific topics include the origin of anatomically modern humans, the peopling of the New World, and the nature of small-scale societies in both modern and ancient contexts. Weekly readings (drawn from the current journal literature), discussions, several short writing assignments.
Note: Required of all concentrators.

THE IDEA IN MORE DETAIL

The most open-ended strategy for designing a K-12 curriculum would be to select instruction blocks one after another solely on the basis of their supposed success in achieving benchmarks (or other agreed-upon specific learning goals). This pure "benchmark-maximizing" strategy would seek simply to hit all benchmarks as often as necessary. Yet, because the blocks themselves carry requirements for prerequisites and resources, the choice of each block would depend to some extent on what blocks have been chosen already and, in turn, would influence subsequent choices. This is a complex demand of design by assembly. Fortunately, it seems likely that there will be computer software for the task (as for almost every task). When proper computer software for curriculum design is developed, it can be used to display a running account during selection of how well the full range of benchmarks is being targeted. Even better would be software that could also identify at each step the next blocks that would most improve the overall benchmarks profile.

Project 2061 is attempting to develop a prototype of such software.

Computer-assisted design by assembly requires three specific and essential resources. First, there has to be a set of related information banks—or, as they are called at this turn of the century, databases:

- A database of learning goals and the connections among them. This database would include *Science for All Americans, Benchmarks for Science Literacy,* the national standards and frameworks for the various subject domains, and state frameworks. It would also include the comparisons in *Resources for Science Literacy* that show the connections of *Benchmarks* to national standards in science, mathematics, technology, and social studies (and eventually to the standards for other subject-matter domains).
- A database containing descriptions of a large number—certainly hundreds and perhaps thousands—of commercially available curriculum building blocks. Some proportion of these instructional blocks should aim at specific learning goals, and critical evaluations should be available to indicate that they are instructionally effective. Equally important, the specific goals, effectiveness, and other properties of each block must be described honestly and well.
- A database of curriculum-design concepts, including some fairly elaborate ones that provide considerable guidance for design, along with exemplary designs that have already been worked out in detail.

Second, there has to be user-friendly computer software available to search these large and complex databases and keep track of options and decisions. Third, there have to be qualified design participants—whether teachers, administrators, school-board members, or citizens—who understand the process of curriculum design by assembly and have the skills and resources to implement it. With the necessary databases, software, and expertise at hand, the design process would consist of five steps:

Step 1: Work toward agreement on a coherent set of specific learning goals to be reached by the K-12 curriculum. The learning-goals database should help here. Avoid taking a cafeteria approach to selecting specific learning goals. The goals should be a related set like those in *Benchmarks* and other major sources. Substitutions can certainly be made, but how they affect the rest of the set needs to be taken into account.

Step 2: Identify and record constraints. Curriculum design is always constrained by policies, resources, schedules, and so forth. It is important to ferret out those constraints, whether explicit or not, that appear to be the most limiting. Of course, all the design principles related to benefits, risks, and trade-offs would apply here.

Step 3: Come up with a design concept that captures what the curriculum should be like or that at least suggests what its main features ought to be. Although different design concepts could address a given set of learning outcomes, some may seem more effective or more motivating than others and also more likely to gain wide acceptance. Scan the database of curriculum-design concepts for ideas. Record the selected or invented design concept, indicating its implications for selecting blocks.

Step 4. Look in the database of completed curriculum designs to see if there are already worked-out K-12 curricula designs that are consistent with the selected design concept and that prescribe an actual array of blocks. If a suitable design is found, then print out the details and begin collecting the specified blocks. If not, use the method described in Step 5.

Step 5: Guided by the design concept, begin selecting blocks. Start by selecting a few blocks that everyone can easily agree upon. Now add more candidate blocks that have some appeal, one at a time. For each block, consider how the goals it targets compare to the slate of goals that are still missing from the collection chosen so far (including specific learning goals and others deemed desirable by the design

Designs on Disk demonstrates a utility for dragging blocks from a database into a curriculum "space" and keeping track of the benchmarks that have been collectively targeted.

Ideally, curriculum-block descriptions should provide information in these categories:

Title

Overview

    Intended Students

    Subject Area

    Format

    Time Frame

    Prerequisites

    Rationale

Content

    Stated Goals

    Main Topics

    Activities

    Options

    Connections

Operation

    Human Resources

    Material Resources

    Assessment

    Teacher Preparation

    Cost

Credibility

    Empirical Evaluation

    Benchmark Analysis

    Reviews

    Users

    Development

Although most of these categories are familiar, their content is sometimes novel. See CHAPTER 4: CURRICULUM BLOCKS for more details.

concept). Keep a record of deliberate trade-offs made or considered along the way, so that both the participants and other people can follow the reasoning, especially if decisions are later to be revisited. Continue adding blocks until the design is complete. Early in the process, choices can be fairly free, because there is so much curriculum space open and so many goals to be targeted. But as the curriculum space fills up, meticulous attention has to be paid to the time requirements of additional blocks and how well the blocks contribute to the still-unsatisfied goals.

Searching, keeping track, and documenting these five steps can be daunting tasks. It may be possible to do the job with pencil and paper, but it would be vastly easier to undertake by using sophisticated and user-friendly computer software. Such software could be helpful in Step 1, searching the goals database and comparing different sets goal by goal. Searches would be facilitated also for design concepts and for complete designs (Steps 3 and 4). To facilitate keeping track and sharing work among participants, files would automatically be set up for considerations made and conclusions reached on local goals (Step 1), constraints (Step 2), and design concepts (Step 3).

It is in Step 5, however, that the greatest benefits of computer assistance would probably be realized. As a new block is tentatively added to the growing array, the computer would compare the goals it targets well to the goals still unsatisfied by the collection. The computer would search for candidate blocks, perhaps even suggest blocks that would make needed contributions to the emerging curriculum. By keeping track also of the resources that selected blocks require, the computer could provide advice on whether candidate blocks meet constraints that had been entered. The computer could even advise on cost-effectiveness—comparing a candidate block's contribution to goals with its demands on time (students' and teachers'), facilities, equipment, and money.

The crowded endgame, when there is too little space in the curriculum and too many unsatisfied goals left, would benefit in particular from computer searching and fitting. Even with computer software to display a running account of targeted and still-to-be-targeted benchmarks, a stepwise series of choices could still lead to a situation in which no set of available blocks that would serve the remaining benchmarks would also fit in the time still available in the schedule. So, as many people experience in packing a suitcase or van, some backtracking may be necessary. Revising some earlier choices would be facilitated immensely by having the computer juggle and compare multiple alternatives and justifications. Curriculum design by assembly as it is envisaged here hardly seems practical without the use of sophisticated computer programs.

SETTING THE STAGE

The curriculum-design system proposed in this chapter is intended to enable educators to design alternative K-12 curricula, ranging from the traditional to the radical, that will all lead to the same desired goals for student learning. The previous section sketched some technical resources that would be required for carrying out computer-assisted design. There are, however, other related conditions, as discussed in the next three sections, that have to be met as well:

- There is widespread agreement among teachers, other educators, and the public on the need for reform, the character of learning goals, the acceptability of curriculum diversity from district to district, and other issues.
- The content knowledge and craft skills of beginning teachers and administrators are far more extensive and sophisticated than was previously the case—and teachers and administrators alike expect to build their skills systematically over their careers. This would imply that the professional preparation of teachers has been thoroughly reformed to be targeted and coherent, rather than piecemeal, hit-or-miss, or driven by instructional fads.
- Educational policies have been adjusted as necessary to support or at least permit this new approach to curriculum design. Accordingly, such curriculum-related matters as assessment, graduation requirements, licensing, the locus and limits of decisionmaking, and funding have been changed where necessary. Perhaps most important, school districts have found a way to make much more time available to enable teachers and administrators to engage in creative and essential activities such as curriculum design.

A dozen aspects of the education system important to curriculum reform are explored in *Blueprints for Reform,* which is available in print and on the Project 2061 Web site at www.project2061.org.

Shared Beliefs

Widespread public readiness for daring curriculum changes designed to improve student learning is not sufficient to guarantee successful reform. There must also be a consensus on what students are intended to know and be able to do—and the recognition that students are not achieving those goals now. At the most basic level, this requires agreement on the balance among the several subject-matter domains and, within each domain, on the balance between the common core of studies and the electives available for students having special needs, interests, and talents. Specifications of the core for the domain of natural-science education are provided by *National Science Education Standards* and *Benchmarks for Science Literacy* (and the strength of those specifications is underlined by

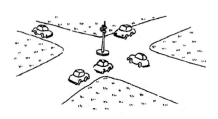

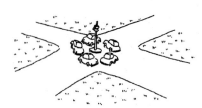

their being almost completely consistent with one another). *Benchmarks* also contains specifications for social science, mathematics, and technology. Although there is not yet a coherent picture of what students should learn across the entire spectrum of subjects, preliminary efforts are under way to link the standards in a rational whole.

One of the premises of the curriculum-design system proposed here is that there is no one best way for students to learn. Thus, different school districts may have very different curricula, despite having the same set of learning goals: common ends, diverse means. The notion of curriculum diversity among and within school districts is not necessarily attractive to everyone. Diversity may be resisted by those who believe that there is one inherently best curriculum design or those who believe that a common curriculum, however imperfect, is necessary to accommodate student migration or varying college-admission requirements. However, where curriculum diversity is tolerated or actively sought by communities, the need for a design system to create serious curriculum architecture becomes clear.

Professional Development

In addition to professional development aimed at building curriculum-design skills, teachers will need opportunities to strengthen and expand their instructional skills. The connection between curriculum design and instruction points both ways. A curriculum sets possibilities and limits on what teachers can accomplish, and teachers determine the degree to which a curriculum's possibilities can be realized. If a curriculum changes radically, so must teachers; if teachers change radically, so should the curriculum. A mismatch between teacher capabilities and curriculum capabilities is a formula for failure.

Effective use of the building blocks of the curriculum of the future will require teachers whose preparation in content is different in breadth and depth from today's norm. Consider, for instance, the content demands of blocks that will be chosen to reach the learning goals recommended in *Benchmarks*: the nature of science, mathematics, and technology and their interconnections; key concepts from the biological, physical, earth, and space sciences; scientific insights concerning human society and the mathematical and designed worlds created by the human species; historical and thematic perspectives that cut across and connect science, mathematics, and technology; and the possession of certain scientific habits of thinking and doing. Those blocks will require that teachers possess knowledge surpassing the level of science literacy outlined in *Science for All Americans*— which may imply extended professional development throughout teachers' careers.

On the other hand, the specificity of science literacy, as defined by Project 2061, makes it possible to focus teachers' own science learning. An elementary-school teacher of science, for example, would not be obliged as an undergraduate to take general courses in biology, chemistry, physics, earth sciences—and, in the Project 2061 notion of "science," courses in engineering, mathematics, history and philosophy of science as well—in the hope that they could sift out and retain the basic literacy ideas. Rather, undergraduate study could be tailored to focus on those basic ideas. (The prescription is made here for prospective teachers in particular, but most undergraduates may best be served in the same way. And some undergraduates may decide only later to be teachers.)

High-quality curriculum blocks could also make teacher preparation more efficient. Say, for example, that the intent is to integrate mathematics and natural science for a semester in middle school. An integrative block that has been developed thoughtfully with ample resources in time and expertise (and that directs the teachers to the particular background knowledge they need) would relieve teachers of having to become broad-based experts in mathematics and science generally.

There are also craft skills to be mastered. In the future, teachers will have to enter the profession with teaching skills that are more sophisticated and more diverse than those that have until recently been considered sufficient. Being able to pace students briskly through a monolithic textbook will no longer suffice in an age rich in multimedia materials, well-developed individual and group methods of instruction, sophisticated assessment approaches, calculators and computers, information networks and multimedia, and steadily rising expectations.

In modern professions, progress seems to lead to and be a consequence of specialization. Teaching has a long history of specialization by grade level and, in the upper grades, by subject matter. It may be that content specialization in some fields like science and music ought to be introduced earlier, but that need not be the end of it. To take full advantage of the curricula of the future may well require the services of teachers who have developed teaching or technical competence of one kind or another beyond that of general practitioners, which they all share. The issues involved in specialization are admittedly many and subtle, but the explicit attention to requisites that good block descriptions will provide ought at least to bring more clarity to the debate.

Education Policies

If different school districts are to take advantage of a resource system that enables them to design different curricula, then state and local policies must permit them to do so. The very essence of design is to find an imaginative accommodation between goals and constraints, making trade-offs between them as necessary, modifying one or

For a list of recommended trade books selected for their focus on basic literacy, see *Resources for Science Literacy: Professional Development.*

"Locating the Vanishing Point"

the other or both in the process. Here are some examples of curriculum-related policy questions that are especially relevant to the curriculum-design system being proposed:

- Where are decisions made? How much freedom does a school district have to design its curriculum? Are learning goals set by the state in detail, in general, or not at all? Are they required or only recommended? Who decides whether a given curriculum design is acceptable—the local school district or the state? And on what basis?

- What regulations are there to ensure accurate and valid descriptions of curriculum blocks? What evidence is required that a proposed curriculum block is likely to result in the claimed learning? For curricula in place, when are assessments required to show that learning goals are being met? What assessment techniques and instruments must be used?

- Do state and local policies (fiscal and operational) permit the adoption of a curriculum design in which some faculty members specialize in research and development, curriculum management, or assessment, and have limited teaching assignments?

- What latitude is there to include blocks that call for teaching by people who are not licensed K-12 teachers, such as students who teach younger students, or scientists from the community who serve as associate or adjunct teachers on a part-time or limited basis?

- Can students receive credit toward graduation by examination in lieu of taking specified courses?

Clearly, there are certain answers to such policy questions that would make it difficult for curriculum-design teams to come up with designs very different from what is now traditional. Other answers would be conducive to the design of more inventively effective curricula and to creating a design-by-assembly system.

Even if these three conditions discussed above were to be met, the proposed curriculum-design system would also require a more reliable and relevant body of high-quality research on which to base curriculum-related decisions. In addition, strong demand for curriculum building blocks would be needed to encourage developers to create many new blocks, describe them fully and objectively, and subject them to field-testing for workability and effectiveness.

ASSEMBLY STRATEGIES

As the five steps proposed earlier make clear, curriculum design in the future would be more efficient, more soundly based, more congenial, and ultimately more successful if undertaken with the assistance of electronic databases and specialized software.

The Idea of Computer-Aided Design

These days, if we want to build our own kitchen, garden, or boat, we are likely to do one of two things. We could search for a ready-made design that seems to be what we have in mind. Alternatively, we could buy a computer program to guide us through the process of creating our own design. Using computers for such design projects is relatively new, although they have been used in industrial and many other contexts since the 1970s.

In education, computers are used mostly for record keeping and student instruction. Both of these applications are important, but both have a long way to go before their potential is fully realized and they become integral to the education enterprise, rather than hit-or-miss add-ons. But computers are rarely used for helping to create and operate curricula.

Different components that could be part of a computer-aided curriculum design system can be found in *Benchmarks on Disk, Resources for Science Literacy: Professional Development,* and *Designs on Disk.* The chief purpose here is to describe how a full-fledged computer-based system could be used in curriculum design, but some attention also is paid to the system's potential use in curriculum resource management, curriculum operation, and curriculum-related professional development.

One caution: Computers and computer programs cannot replace human inventiveness and vision. No matter what instruments are used—legal pads and #2 pencils or sophisticated computer programs—the curricula that are created can be no better than the thinking that goes into them. The subject here is computer-*aided*, not computer-*generated*, curriculum design and management.

Searching for Candidate Blocks

In the design-your-own-sound-system analogy used earlier, a person shops in stores or catalogs, or both, to find out what is

CAD/CAM

One of the first large-scale uses of computers in industry was in designing products and systems and guiding manufacturing operations. Hence the acronym CAD/CAM: Computer-Aided Design/Computer-Aided Manufacturing.

The construction and assembly of products is a complex undertaking, from the ordering and flow of raw materials or parts to the sequence of connecting them. How it is carried out can make a great difference in the efficiency and cost of production.

Manufacturing considerations also have implications for the design of products, since small changes in design may also make large differences in the efficiency and cost of construction. The design of products and the design of constructing and assembling them are therefore closely linked. It has turned out that CAD/CAM has useful applications in a wide variety of contexts, including the design of research investigations, traffic patterns, buildings, computer chips and computers, magazine layouts, weapons, parks—and in other situations that require extensive data to draw on, careful tracking, and a variety of alternative models to explore.

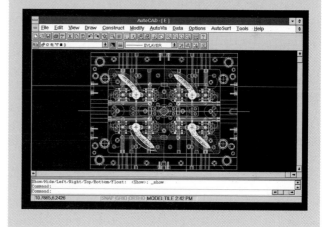

Some day there could be an on-line "curriculum store" that includes not only the block descriptions asked for in Chapter 4, but also video demonstrations of what blocks look like in action.

available and to compare the options in terms of expected performance, appearance, and cost. Then, taking certain constraints into account (such as budget and space limitations) and applying certain selection principles (such as which kinds of components are desirable and compatible), selections are made. In that analogy, on-line guided selection could have substituted for the store and catalog.

There is, of course, no such thing yet as a comprehensive "curriculum store" where one can examine an array of curriculum blocks critically and get precise information on their properties and costs. The jumble of materials on display at education conventions and on the Internet, even the displays of the major textbook publishers, provide very little information of use to curriculum designers. But if one assumes that in time many hundreds of good curriculum blocks will exist, the question arises, How can we possibly deal with such a vast amount of information?

A design team (whose membership might differ greatly from one district to another) begins a search for suitable blocks by selecting the categories of variables it wishes to have taken into account in the search. The box on the opposite page, A Variety of Block Shapes, illustrates a wide range of temporal dimensions of blocks. A template for essential block information is described in detail in CHAPTER 4: CURRICULUM BLOCKS.

Although a computer makes it possible to search on the basis of any of the variables in the block description, an efficient search technique is to build a pool of candidate blocks based on overview properties. Using the AND's, OR's, and NOT's of search logic, the team uses the curriculum-design software to have the computer find all of the blocks that meet its audience, subject-matter, coherence, format, and time requirements. Here are some examples of block properties a team could ask for:

- Project blocks emphasizing observation and collection by K-2 students
- Any blocks with especially rich options for advanced students
- One-semester integrated science/mathematics blocks organized primarily around measurement for all students in grades 3-5
- Quarter-long, discipline-based, grade 6-8 mathematics core courses featuring the use of statistics in demographic and economic applications
- Any blocks that claim to target goals from CHAPTER 11: COMMON THEMES of *Benchmarks for Science Literacy*
- One-semester integrated science/history seminars for all grade 9-12 students
- Any blocks for which there is empirical evidence of student learning
- Year-long courses in calculus for students in grade 12 who plan to major in science and mathematics

A VARIETY OF BLOCK SHAPES

Because of the demand of filling 13 years of curriculum, and the desire for more coherent instruction, blocks are likely to be fairly large units, certainly larger than "activities" and probably larger than most "units"—probably more on the scale of courses or half-courses. In this scheme, a block's length represents its calendar duration and its height represents what share of time it gets on the scale of minutes per day or days per week. A conventional high-school course would correspond to a block with a "width" of a school year and a "height" of about 20 percent of instructional time (one period a day, five days a week):

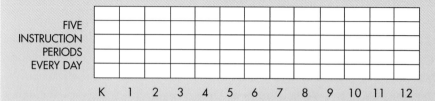

A curriculum made up of nothing but such courses (call them subjects in the lower school) would look like a wall of uniform bricks:

FIVE
INSTRUCTION
PERIODS
EVERY DAY

K   1   2   3   4   5   6   7   8   9   10   11   12

But there are many other possibilities. Even keeping the same total hours, a block might be greatly compressed or extended: at one extreme, an experiment block could fit into a month of all-day sessions; at the other extreme, a seminar block could require only one period per week for five years. Blocks could be still more different in "shape," say a weather-cycle block that would take only five minutes a day over several years, or a course-sequence block that would occupy all of the middle-school years. These possibilities would correspond to the block shapes drawn below.

Some extreme examples of possible instructional blocks:

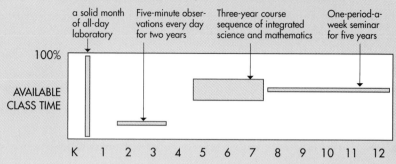

Using blocks with a variety of shapes obviously might cause severe scheduling problems, unless the blocks and the school are suited to "modular" scheduling, in both daily and calendar time dimensions.

There already exists curriculum-accounting software that helps to lay out chunks of time for every element of the curriculum framework. What happens in those chunks, and whether they are enough, is another matter.

A search of the curriculum-block database will turn up only those blocks having the requested combination of properties, displaying the titles of candidate blocks. If the search turns up too few or none, some of the constraining specifications may have to be removed from the search request. If a daunting number of blocks turn up, more constraining criteria may have to be added. Clicking on a title in the database brings up a brief summary, the equivalent of a college-course catalog description. On the basis of the summaries, some blocks can be eliminated from further consideration, thereby reducing the size of the candidate pool.

Configuring Curriculum Blocks

Choosing among candidate blocks to be considered for inclusion in the curriculum involves making two kinds of decisions. First, each block in the candidate pool must be evaluated on its individual merit. How many specific learning goals does it target? How appealing is its instructional strategy? How does it compare with its competitors? Does it fit the overall design concept? (The box opposite illustrates how the same design concept could take different forms.) Second, since the block will end up as only one part of the curriculum, it must be evaluated in light of its special contribution to the whole array of chosen blocks. Does it target goals still unserved in the collection of already chosen blocks? Does its format provide needed balance to the different formats in other blocks? Does it supply timely prerequisites for later blocks, and are its own prerequisites met by earlier blocks?

As the configuration grows, serious analysis of profiles, patterns, and trade-offs can begin. Are some goals targeted again and again while others are ignored? Is the redundancy needed or wasteful? How difficult is it to find blocks to fill the gap? Are we getting the balance we want among block types? Where are we weak? Are there slots for which better candidates should be found? What if we added these two blocks in place of that one? What would be the consequence for the block's own grade level and for those it connects to? And so forth.

As this analysis proceeds, the design team can follow up on references and other elements of a block description. References to published reports, research, or reviews that support them (and possibly the documents themselves) can be obtained through the Internet and then studied as decisions are being reached. The block descriptions also cite schools or districts where such blocks are in use, so members of the design team can talk with or even visit educators who have experience with a block in which the team is interested. As decisions are made, a satisfactory design finally will emerge, but keep in mind

A design concept can be realized in a variety of different patterns, some applying to *every* block, some to the whole pattern of blocks. For example, a design concept that focuses on using a particular instructional format 60 percent of the time could specify that 60 percent should use only that format or that a certain *proportion* of the blocks should use only that format. Alternatively, the concept could specify that there should be a certain *proportion* of that format within each block or that there should be a balanced *variety* of instructional formats, either across or within blocks.

Three ways in which a distinctive curriculum feature **X** could be distributed among blocks to achieve 60 percent **X**:

Pure format X in 60% of blocks:

60% format X within every block:

Some of each:

that in design "satisfactory" means "adequate for the purpose," not "perfect."

This process sounds long and complicated in the telling, but there is nothing about it that educators could not do using today's computers, if only the blocks existed. The diagram that follows illustrates the kind of information that a computer-assisted block search could present to the designer as a candidate block is added to a partially filled curriculum design space. When an icon for a new block is dragged into an unscheduled opening in the design, its dimensions show how much time the block would take, and bar graphs show how many new benchmarks it would target.

The diagram, hypothetical as it is, shows only two kinds of blocks—one period a day for one or two years. A much greater variety would be possible, as illustrated in the box on page 115, which offers an impression of the diverse array of blocks that could fit into just the 6-8 grade range. The result of applying this process is a curriculum that has been thought out top to bottom by the educators who are going to implement it and by the community it will serve. Nevertheless, the parts will have been created by experts with time and resources unavailable to teachers.

In configuring the curriculum blocks, the selection inevitably gets more difficult as the curriculum space fills up. Blocks to fill in missing goals may not be found or may not fit in the remaining time. Learning goals still left to target could be from scattered content domains. (For example, the design team could end up needing a single

The consequences of adding a block to the curriculum might be displayed this way with block-assembly software.

BLOCK-ASSEMBLY SOFTWARE

Dragging and dropping a rectangle that represents a candidate block produces a display of bar graphs showing how many still-untargeted benchmarks the block aims at for each chapter of *Benchmarks for Science Literacy*. The bar graphs also show how many total benchmarks there are for each chapter and how many have been targeted by other blocks already selected. Additional block characteristics can also be displayed. If the candidate block is added, the benchmarks account is updated. Even if the block is not selected, a record of the attempt can be stored for later review.

Blocks in curriculum design

Number of benchmarks

☐ in *Benchmarks* chapter
☐ targeted so far
■ targeted in candidate block

CHAPTERS

1: The Nature of Science
2: The Nature of Mathematics
3: The Nature of Technology
4: The Physical Setting
5: The Living Environment
6: The Human Organism
7: Human Society
8: The Designed World
9: The World of Mathematics
10: Historical Perspectives
11: Common Themes
12: Habits of Mind

Candidate block

☐ OK to add
☐ Get more information
☐ Don't add, but log idea
☐ Cancel addition

Display other properties

☐ Core/elective balance
☐ Instructional formats
☐ Discipline/integrated balance

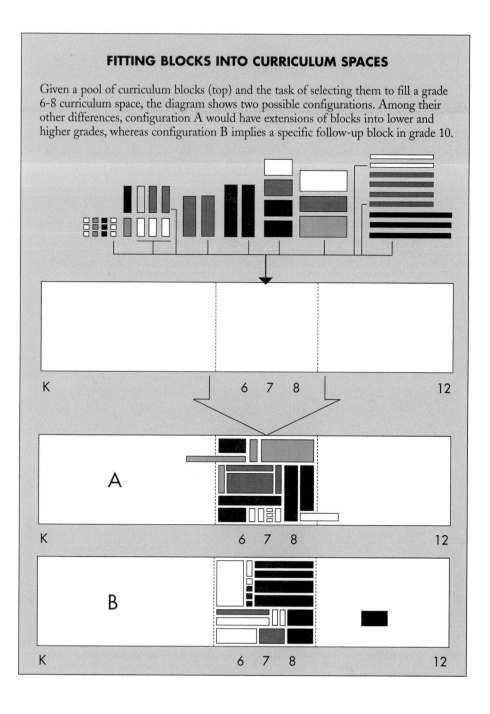

FITTING BLOCKS INTO CURRICULUM SPACES

Given a pool of curriculum blocks (top) and the task of selecting them to fill a grade 6-8 curriculum space, the diagram shows two possible configurations. Among their other differences, configuration A would have extensions of blocks into lower and higher grades, whereas configuration B implies a specific follow-up block in grade 10.

final block that targets benchmarks from the nature of the scientific enterprise, agri-cultural technology, relativity, and cells.) Furthermore, the remaining benchmarks could be remaining, not just because of the luck of the draw, but because they are par-ticularly difficult to learn or otherwise troublesome to develop instruction for.

It is entirely possible that no acceptable configuration of blocks will satisfy all of the goals in the allotted 13 school years of 180 days each. One reaction to that out-come would be to go back and identify some blocks that were selected for reasons other than their effectiveness in serving specific learning goals and to replace them with blocks that are more targeted to benchmarks. Other possibilities could be to lower the priority of some learning goals, squeeze more blocks into the same time frame, or shift the balance of time for different subjects within the school day. A more extreme reaction might be to lengthen the school day or year. On the other hand, it is also possible, if block development has really been successful, that there would be time left over. Depending on where the free time shows up, some school districts could choose to fill out the last years with electives, whereas others could choose to spread the core curriculum out more thinly over all 13 years and provide electives each year.

A Nutritional Analogy

We can get an idea of the kind of help computer-assisted curriculum design could pro-vide by considering an existing example of design-by-assembly software from outside education. The next box is based on an on-line analysis program for selecting a day's menu of food items. The nutrition-analysis software tabulates a cumulative profile of nutrients for the food items that have already been selected. But it is the user's respon-sibility to scan the profile and notice where it still falls short of recommended nutrient totals. Having noticed a shortfall in, say, calcium, the user can request the program to display of a list of food items that are high in calcium—and then choose one from among those candidates to add to the menu. A more helpful computer program would be one that can point out shortfalls itself, then display a list of candidate food items that would contribute well to filling in all of the missing nutrients (while not greatly exceeding other requirements that were already satisfied). The user, of course, would still make choices among those candidates according to individual preference.

One can even imagine that the program could be used to proceed step by step to design a complete diet on its own. In its simplest form, that would mean first choos-ing the single food item that would satisfy the most nutrient requirements, then choosing the next item to best satisfy the remaining requirements, and so on. If the

AN EXAMPLE OF COMPUTER-ASSISTED DESIGN: DAILY DIET

On-Line Nutrition Analysis Tool

Step 1. The dietary suggestions for men, women, and various age groups are different. Please select your appropriate age and gender categories.

Step 2. In the accompanying list of the most common nutrients in the USDA database, click on any nutrient for an explanation of what it is, what it does in the body, and some common foods in which the nutrient is found. Click on each nutrient you want to include in the analysis.

Step 3. Add foods to your personal diet list. From the lists provided, click on a food name, a particular form of that food, and the number and size of servings. The list will keep track of this information and also the total weight of that food.

Step 4. Inspect the nutrient analysis table to see an account of the nutrients in each food choice and the total for the entire daily diet selected so far. The table will show both the amount of nutrient and what proportion that amount is of the recommended daily total.

Step 5. Identify nutrients that are still undersupplied and click on Suggested Foods to see list of foods rich in those nutrients. Select new food items to add to diet list (or increase amounts of some food items already selected that are rich in that nutrient).

Example table:

Food Item	Serving Size	Servings	Calories	Protein	Fat	Carbo-hydrates	Vitamin C	Calcium
Porterhouse Steak, Choice	4 oz.	2	692	56.3g	50.1g	0 g	0 mg	18.14 mg
Potato– baked with sour cream, chives	1 potato	1	393	6.6g	22.4 g	50.1 g	33.8 mg	106 mg
Apple–raw	1 apple	2	81	0.3 g	0.6 g	21.1 g	7.9 mg	9.7 mg
Daily Total			1166	63.2g	73.1g	71.2 g	41.7 mg	133.8 mg
Daily Recommended for female aged 25-50			2200	50 g	73.3 g	---	60 mg	800 mg
Percent of Daily Recommended			53%	126.4%	99.7%	---	69.5%	16.7%

Adapted from *Nutritional Analysis Tool, v 1.1* (http://spectre.ag.uiuc.edu/~food-lab/nat/) Department of Food Science and Human Nutrition, University of Illinois at Urbana-Champaign.

food-item database included additional variables such as the weight and cost of food items, the program might be able to design diets that would be optimally lightweight or cheap, while still satisfying nutrient requirements. (Instead of favoring food items that provide the most nutrients, it might favor those that provide the highest ratio of nutrients to weight or nutrients to cost.) There is no guarantee, of course, that such an optimum diet would be palatable to any particular user (or to anyone at all). Desirability would require user preferences to be invoked at each step—say, no broccoli. Straightforward user preferences that could be clearly specified (for example, for vegetarianism or food allergies) could also be entered and taken into account by the program, but it seems likely that many preferences would be subtle and interdependent, always requiring user involvement to get satisfactory diet designs.

The analogy to curriculum design is fairly obvious, with nutrient requirements becoming specific learning goals and food items becoming curriculum blocks. Some user preferences that could be readily specified could be entered into the block-selection program, whereas others would require close involvement of the users in choosing among candidate blocks found by the program. The total cost in material and human resources would be an important variable to keep track of, as would the total time. (The block that targets the most benchmarks, even with demonstrated success, may also take up a great deal of curriculum time.) And, given the severely limited time frame for the curriculum, the raw number of benchmarks targeted might not be as important to consider as the ratio of benchmarks to the time required for it. Curricula designed this way can be described as "benchmark-efficient" or "learning-optimizing" curricula—although at the cost of not having any other unifying character.

CURRICULUM-RESOURCE MANAGEMENT

The design-by-assembly process will result in a curriculum design, but such a design is not a curriculum. Clearly, implementation of the design to create an actual curriculum that works as intended requires appropriate human and material resources. Computers can aid in meeting those requirements as well.

Material Resources

Today, much of the burden of locating resources falls on teachers. If the subjects or courses to be taught are based on a textbook, suggestions on resources may be given in the accompanying teacher guide. That helps. Often, however, the guidance is scanty

The discussion of specific learning goals here is presented in terms of "benchmarks." In principle, any coherent, progressive set of specific learning goals can serve as benchmarks. Project 2061 tools include utilities for translating other sets of national, state, or local goals into benchmarks, which may serve as a common currency for curriculum analysis.

and limited to apparatus and kits supplied with the textbook—databases, computer programs, and so on. On the other hand, sometimes the teacher guide is huge, with information on myriad extra (but not always relevant) activities—to the point of discouraging teachers from using it. Thick or thin, however, the amount of material is not so important as how explicitly and well it is tied to specific learning goals.

The process used to create a curriculum design can also be used to identify the resources of all kinds needed to implement the curriculum. The block-description templates list all of the materials needed for each block. As a result, the computer can keep track of the total resource requirements (and cost consequences) as each new block is added to the configuration and the design unfolds.

Once the design is complete and has been approved, computers can be used to help organize, monitor, and operate the business aspects of resource management. Accounting software will coordinate with block-selection software to provide a summary list of all the materials needed, when, in what quantities, and at what estimated cost. If major budget constraints loom, the design team can recommend alternative blocks.

The curriculum-design approach sketched here calls for a much greater diversity of learning materials than in the past, and the materials themselves are less likely to be aggregated or even available on-site. The Internet seems likely to become an increasingly important instructional resource, and a way will have to be found to apportion access to it fairly. In the future, school-district computer centers will be able to deliver instructional materials to the classroom on demand or just in time. With such complexity, computer assistance will be essential.

So much management responsibility can surely seem intimidating to teachers today, but remember that we are imagining situations 10 or 20 years in the future. By then, teachers will have grown up with the kind of technology we are talking about and using it to plan instruction will be a key part of their professional preparation.

Even if all that occurs, teachers will be responsible for managing the process wisely. To that end, they will need analytical tools, probably the more integrated the better. The analytical and record-keeping parts of *Designs on Disk* will become part of the curriculum CAD/CAM system. (Here "CAM" stands for "Computer-Aided Management.") As a result, teachers will be able to examine new materials critically and enter their findings in the resource database. That will allow them and others to update and refine the resource inventory, substituting better materials as they come along. This electronic "review journal" will encourage teachers to share their analyses and conduct discussions about them on-line.

Human Resources

Curriculum resources are not limited to materials. The single most important resource, no matter what the curriculum, is the teacher. In curricula of the future, teachers will continue to be the most important resource, but their roles, techniques, and training will change. And as the curriculum, instructional materials, and teachers change, so too will the management of students.

Diversified teaching roles. As the curriculum becomes more complex, teaching roles will diversify. Think of a curriculum made up of courses, seminars, projects, and independent study—some plainly discipline oriented, some integrated to a greater or lesser extent. In these blocks, time may be configured in many different ways, and distinctions between grades may be blurred. In such a curriculum, different teachers will specialize in different aspects of the work. Some may teach blocks that cut across subject-matter domains, some may teach straight discipline courses, some may specialize in monitoring independent study or organizing and supervising seminars, still others may train and supervise students in peer teaching, and so forth. Keeping track of such a variety of assignments will be much more complicated than in traditional curricula. In addition to having individual skills, of course, teachers will need to be skilled in working in teams in which they make specialized contributions.

This picture will become more complicated as the teaching responsibilities in future curricula become diversified. In addition to a central core of highly trained certificated teachers, future curricula will very likely depend on peer teaching, on the use of adjunct teachers (especially for beyond-the-core blocks), and on access to remote teachers by means of computers and telecommunications (as virtual classrooms become part of the curriculum). To organize and monitor such an enterprise will require computer applications that combine appropriate spreadsheet, database, and design functions.

For some examples of the kinds of functions the curriculum design software will be able to carry out, see Designs on Disk.

Student programs. By the same token, students of the future will follow more varied routes than is now the case. There will be the core program, in which students may or may not follow a common sequence of studies, and the beyond-the-core elements, in which student programs are likely to differ greatly. Moreover, to the degree that a curriculum bases student progress on what students have learned rather than on what they have "taken," record keeping will take on a new dimension. Student programming and student record keeping will necessarily have to become computerized, a trend already under way. We assume that in the dozen or more years before all this is likely to come to pass, administrative policies, teacher preparation and career development, and public expectations and support will rise to the challenge and make it feasible. Such functions should be closely tied to the curriculum, and so an effort is being made by Project 2061 to include those management functions in software it is developing for computer-assisted curriculum design.

CONTINUING PROFESSIONAL DEVELOPMENT

There is no substitute for strong preparation in the content of one's field or for a similarly strong preparation in the techniques of one's craft. This is no less true of teaching than it is of engineering, medicine, or any other advanced field. Throughout one's career, knowledge and skills must be continually updated.

For teachers, preservice preparation can be generic in that it need not be specific to any particular curriculum, although it should include the intimate study of and practice in applying the relevant standards. Then, once actual practice begins, there is much to be gained by focusing professional growth on the demands of the curriculum at hand. In the future, the curriculum building blocks will make it clear what they require to be taught properly and will indicate sources for gaining the requisite knowledge and skills. In addition, the professional development material—including suggestions for workshops and study groups within the district or school, and for carrying out a program of self-directed study—now found on *Resources for Science Literacy: Professional Development* will become part of a complete CAD/CAM system and be continually improved.

See Chapter 6 for a more in-depth discussion of professional development activities that focus on science literacy goals.

Although there will still be a need for teachers to take formal advanced refresher courses from time to time, the bulk of continuing education will take place as part of an individual's regular work. In addition to the guidance that can be provided by a sophisticated, computer-based multimedia system with telecommunications links to information, instruction, and experts, the curriculum itself will have to provide time for professional growth. The traditional preparation period does not accomplish this, and occasional half-day training periods hardly do better. The curriculum structure of the future will open up the way and provide more opportunities for study, and the curriculum blocks themselves will engage the teacher in a way that will make the teacher's own learning integral with, or at least parallel to, that of the students. There will also be other conditions needed for effective professional development to become a reality: an office, a telephone and a computer in one's office, a budget for professional development, and, above all, the establishment of appropriate professional expectations and attitudes.

Calvin and Hobbes   by Bill Watterson

An Egyptian Garden, 1400 B.C.

CHAPTER 4
CURRICULUM BLOCKS

For millennia, designers of plazas, walls, arches, bridges, and domes have used bricks as their basic construction units. Identical bricks can be arranged in a surprising number of different patterns, and many times more patterns can be created using only a few different kinds of brick. Some aspects of the patterns are essential to the purpose of the structure, whereas other aspects may be merely pleasing. To suit one purpose or another, the bricks can be different in material composition (marble, stone, glass, concrete, metal, plastic, clay, wood), structural properties (size, shape, strength, resilience, durability, response to temperature), appearance (color, transparency, surface texture), and inevitably cost (bare minimum to extravagant). How well the structure will do its basic job depends more on some of these properties than others, but any property may be important in motivating and selling the design. Obviously a designer needs to know what the properties of bricks really are and what features the final design must have.

In the 20th century, sad to report, curriculum designers have not been so well blessed. For most of the century, they have had only a few kinds of "bricks" available to them, and those in only two "shapes"—one period, five days a week for one semester and one period, five days a week for one year. To make matters worse, there simply is no agreed-upon set of basic properties to use to describe or talk about curriculum components. For this and other reasons, "honesty in packaging" is rare in the descriptive materials provided by publishers.

This chapter supposes that in creating the curriculum equivalent of a brick structure, educators will have access to a large variety of accurate descriptions of bricks in the form of an easily accessible database. After defining what a curriculum block is, the chapter presents a template for describing blocks, some thoughts on how and by whom suitable blocks can be developed, and a brief list of some possible curriculum blocks.

Examples of brick patterns

WHAT ARE CURRICULUM BLOCKS?

From a psychological point of view, the 13 years of learning in grades K-12 can be regarded as a single fabric of acts and thoughts, each of them occurring in the context of all the previous ones. From an administrative point of view, on the other hand, the 13 years can be described as a sequence of discrete parts—lessons that are grouped into units that are grouped into courses that are grouped into curricula. Thus, for example, a lesson on the cell wall may be found in a unit on cells that is part of a biology course that is part of a high-school curriculum (which is part of a K-12 curriculum).

How Big Is a Curriculum Block?

The lesson —▸ unit —▸ course —▸ curriculum hierarchy is less rigid than it seems, since students' learning experiences are not always shaped into discrete lessons, lessons into units, or units into courses. In any case, for purposes of secondary-school curriculum design, it is best to use courses (or their equivalents) rather than lessons and units as curriculum building blocks. Dealing with smaller components is extremely difficult in view of the huge numbers involved and the myriad connections among them. Also, unlike lessons and units, courses are formally labeled, administrative blocks of learning activities for which students can receive summary credit (and usually report-card grades) with consequences for student promotion, graduation, college admission, and employment.

By the same reasoning, the subjects taught in elementary schools—not usually designated as courses—are also curriculum building blocks. And if we assume, as is done here, that other kinds of curriculum components comparable to traditional courses and subjects in magnitude and coherence are possible, then it makes good design sense to think of curriculum blocks as the large structural components for which students receive separate credit on their transcripts. By this definition, elementary-school subjects and secondary-school courses are curriculum blocks, but the units and lessons that make them up are not.

The term "curriculum block" is short for "curriculum *building* block." It reminds us (to a degree that "subject" and "course" do not) that a curriculum is *constructed* by selecting and configuring a relatively few major components, not hundreds of parts. A curriculum *design*, then, is a plan for their selection and configuration. Correspondingly, blocks must have properties that enable them to be so assembled—and those properties must be evident to curriculum designers.

A Common Block Example

As simple an idea as the curriculum block may seem, it has subtleties that require close attention. Let's consider a traditional block commonly found in the core curriculum of most school districts today. The biology course taken in 9th or 10th grade will do nicely since it is taken by virtually all students. Here is an imaginary discussion between a curious reporter and a high-school curriculum specialist:

Question: How would you describe the biology course in your school or district so that others would know what it is like? Answer: I would name the biology textbook we use.

Question: What specific knowledge and skills are all the students expected to acquire by taking the course? Answer: They will understand the key concepts of biology and what biological science is like. And they will come away with greater respect for nature. This is asserted in the preface or first chapter of every biology textbook.

Question: Those aims seem to be pretty general. Can you be more specific about exactly which key concepts the students are expected to learn and at what level of sophistication? Answer: That would be evident in the chapter and section headings of the textbook—or, better yet, in its glossary and end-of-chapter questions.

Question: Judging from biology textbooks I have seen, students have a lot of material to get through. Are *all* students expected to learn *everything* in the text? Answer: Usually not. Our teachers decide what to leave out, what to treat lightly, and what to bear down on. And of course, there are always some pet topics or activities. So I guess you could say that the textbook doesn't quite define the course content, but mostly it does.

Question: Where can one find a description of how the biology course takes account of the student learning that is expected to precede it and that which will follow? Answer: One can't. Biology is designed to stand on its own because, on the input end, teachers commonly say they cannot count on students entering the course with specified knowledge and, on the output end, many students do not go on to take chemistry, physics, and advanced biology.

Question: Does the content of the textbook you use align with national standards? Answer: Yes, there is a table in the teacher-guide version of the text that indicates which textbook sections correspond to both *Benchmarks for Science Literacy* and *National Science Education Standards*.

Question: But don't publishers routinely use such tables to make superficial connections to "topics," without attention to what the actual knowledge that students are expected to acquire? **Answer:** I suppose that may be true, but we haven't anything like the resources required to make a careful study of idea-by-idea alignment.

Question: If a textbook pretty much defines the course, does it describe how instruction is to be organized and the results evaluated? **Answer:** Yes and no. Mostly that is up to the teacher, who may change the order of topics somewhat. However, in most schools, you will find biology being taught in the traditional format involving assigned reading in the textbook, homework, class discussion, teacher demonstrations, laboratory, and periodic tests. Many teachers also have students do individual projects. Textbooks are accompanied by separate teacher's editions or guides that suggest how to organize instruction. And by tests.

Question: What "space" does biology have in the curriculum here? **Answer:** The same as every other science course, or any other course at all, for that matter, which is to say one period every day for one school year. It is placed in the 10th grade, where it has been stationed for as long as anyone on the faculty can remember.

Question: Finally, how many decidedly different versions of biology courses are generally available to select among? **Answer:** If we take textbooks to define courses, there are about three broad types on the market. Several traditional biology textbooks, and hence courses, are based on an organization of content going back to at least the 1930s; a good part of each of these is classification of organisms. But as biological knowledge has increased, more and more has been added to these textbooks until they are virtually encyclopedias, with very little on any one idea and sparse connections made among ideas. All teachers who use them have to leave out some chapters, and some supplement them with other materials for important topics. (In fact, some teachers are beginning to use the heavy, encyclopedic textbooks only as supplementary reference books.) There are also textbooks with a more modern organization—based, say, on organ systems or habitats rather than phlya, though much of the content is not very different. Then there are one or two that have greatly reduced the coverage in favor of trying to focus on a modest number of important ideas. We hope there will be more of those coming along to consider. So we feel we have a fair number of possibilities to choose from.

From the above imaginary discussion, we could conclude that biology is in relatively good shape from a curriculum-design standpoint. There are significant differences among the biology offerings, not simply cosmetic ones; the offerings have distinct conceptual approaches, and they provide or identify needed materials and procedures, yet leave room for teacher creativity. Still, there are some shortcomings from the design standpoint. It is usually too difficult to get details about alternative courses, except in rather general terms. Textbook-selection committees (the name for course-selection committees) do not have easy access to vital information, such as precisely what knowledge and skills are being targeted, what trade-offs there are among the competing courses, and what evidence there is that each course will achieve what it claims.

Another shortcoming deals with the variety of courses offered. In spite of the thematic differences among the specific biology courses, welcome as they are, it would be helpful to have other possibilities: semester and quarter courses that target national-level literacy goals through a focus on biotechnology or biodiversity or on medical or agricultural applications, or seminars that meet only two hours a week and are based on case studies or readings. Other interesting possibilities are courses centered on a historical episode—say, one using Jonathan Weiner's *The Beak of the Finch* (1994) as a focus for the study of Darwin and his work. Biology blocks might consist only of inquiry projects or computer-assisted independent study spanning several years. Nor is there any compelling reason why general biology in high school should be confined mostly to the 9th or 10th grades. These and similar possibilities for greater variety exist for other science, mathematics, and technology subjects.

PROPERTIES OF CURRICULUM BLOCKS

In short, the notion of curriculum blocks focuses attention on having a wide variety of "bricks" from which curricula can be assembled, and also having precise descriptions of blocks that will enable educators to make informed choices and placements of them in designing curricula. The point is not that every curriculum should have a diverse variety of blocks, but that a diversity of blocks be available so that it is *possible* to incorporate as much or as little variety into a curriculum as a school district wishes in pursuing specific learning outcomes. With that possibility in mind, we can state that

The advent of a national consensus on specific learning goals may make it more feasible and desirable for publishers to describe in detail and in a common language the specific goals that a course targets.

Full-year biology courses could include these variations as components—if the total burden of topics to be covered were drastically reduced.

- **A curriculum block is a self-contained sequence of instruction.** A block is important enough to require school-board approval and to be recorded as an entry on student report cards and transcripts.
- **Different curriculum blocks can have different time dimensions.** They need not be restricted to one period every day for a semester or year. In principle, "courses" could have different time dimensions, but the tradition is otherwise and has proven to be stubborn.
- **A curriculum block can have any of several instructional formats.** There could be seminar blocks, project blocks, independent-study blocks, lecture-series blocks, and peer-teaching blocks, in addition to the traditional course blocks. Although elements of any of these formats can be included in more traditional courses, they may be more useful in their own right than they would be as occasional elements. Any one block may include a deliberate mix of formats, too.
- **A curriculum block can relate to one discipline or several.** The content of blocks may feature the nature and conclusions of a particular discipline or alternatively explore particular phenomena (such as lakes, explosions, transplants) or issues (pollution, space travel, biodiversity), drawing on several disciplines as needed. Mathematics courses customarily have been organized by divisions of the discipline, whereas social-studies courses often are organized around events or issues.
- **A curriculum block is adequately described.** A block description provides curriculum designers with sufficient information to enable them to make informed choices. Two chief items of information are a delineation of the specific learning goals that are credibly targeted by the block and an indication of the degree to which the claimed results have been validated.

A TEMPLATE FOR DESCRIBING CURRICULUM BLOCKS

Block descriptions are needed to judge how a particular curriculum block would fit into a curriculum. The set of descriptors that follows constitutes a template for making entries in a database of block descriptions. Worked out gradually over a period of years by members of the six Project 2061 school-based teams in concert with curriculum specialists and materials developers, the descriptor set is a straightforward formulation of what any group of professional educators would want to know when selecting blocks. The template is summarized in the accompanying box, and each basic component is discussed below. (The discussion includes examples drawn from science,

A TEMPLATE FOR DESCRIBING CURRICULUM BLOCKS

Title

Overview

    Intended Students: grade range and special groups that block is designed for

    Subject Area: areas of study (e.g., chemistry, geometry, social studies)

    Format: how instruction is organized (e.g., traditional course, seminar, independent study)

    Time Frame: options for calendar duration, meetings/week, and time/meeting

    Prerequisites: prior knowledge and skills needed for student success in block

    Rationale: ostensible purpose for study (e.g., explain phenomena, follow a social issue, design a product)

Content

    Stated Goals: specific learning outcomes that developer claims block serves

    Main Topics: what will be studied (e.g., nutrition, recycling, maps)

    Activities: typical student experiences (e.g., measuring rainfall, debating policy)

    Options: alternative materials, activities, or organization for teachers or students

    Connections: relevance to subsequent or parallel parts of the curriculum

Operation

    Human Resources: professional or lay staffing needed for instruction or support

    Material Resources: needed equipment, supplies, sites, and transportation

    Assessment: tasks, scoring guides, and schedules for assessing student progress

    Teacher Preparation: needed knowledge and skills in subject and teaching

    Cost: time and money needed

Credibility

    Empirical Evaluation: scientific studies of what students actually learn from block

    Benchmark Analysis: evidence-based estimate of which benchmarks are served*

    Reviews: published expert opinion on the quality of the block

    Users: places (and contact people) where block has been implemented

    Development: how, when, and by whom block was created

* Benchmark analysis is described in CHAPTER 6: BUILDING PROFESSIONAL CAPABILITY.

mathematics, and technology, but, with minor changes in wording, the template can be made to fit any curriculum domain.)

The Project 2061 template for curriculum-block descriptions was created to serve as a standard for a national database of such descriptions. The task of building a complete and valid database needs to be national in scope, because the resources needed to acquire and evaluate a large number of curriculum blocks probably are beyond the means of local or even state groups. Although some of the description categories called for by the template are familiar and routine, others—such as empirical studies and benchmark analysis—are novel and demanding. And even some of the seemingly routine categories have subtle spins that merit fresh attention. Sometimes there will be considerable redundancy among the title, subject area, rationale, main topics, and activities of a curriculum block. A title as straightforward as "Learning about Motion through Catapult Design" pretty much covers all of these descriptions. More often, the activities in a block may not be easily inferred from its title or from any of the other description categories. For example, a template could look like this:

Title: Catapult!
Subject Area: physics of motion
Rationale: design a projectile device
Main Topics: elasticity, trajectories
Activities: sketching schematic diagrams, testing materials, piloting and revising designs, and distance contests

Different as they are, each category conveys something helpful to know about the block. The organization of information shown in the template is not meant to be the only one possible—and certainly does not imply importance or priority. From the Project 2061 point of view, the last category, Credibility—particularly Benchmark Analysis—is probably the most important, for that is where the focus is sharpest on what students are likely to learn. Which information designers would look at first in searching for blocks would, of course, depend on their search strategy—and probably on how far along in their search they were. For example, Prerequisites might not be an important concern early in selecting blocks, but will become increasingly important as the curriculum space fills up.

It should be kept in mind that the information in the database is the same information we would want about the courses making up the existing curriculum in a school district. It is sometimes an eye-opener to discover how little—if anything—is known about the components of an existing curriculum or, how little is systematically recorded and readily available.

Title

Curriculum designers need convenient labels for referring to each block they consider. They can, of course, make up any labels they wish, but it would be helpful to have a convenient title offered by the developer or database maker. The titles do not have to be uniquely distinctive, so there is no need for a national registry, such as there is for the many millions of site names on the Internet. If publishers X and Y both put out a textbook called *Modern Earth Science*, designers would have no trouble referring to "X's earth science" and "Y's earth science." Similarly, one could easily refer to a hypothetical "Philadelphia Schools' environment seminar" or "Livermore Labs' environment seminar." Although it is always helpful for each title used in the database to reveal something about the block's contents (for example, BSCS's "Science and Health Sequence"), the titles of some highly respected blocks have not even hinted at what was to be learned (for example, Bank Street's "Voyage of the Mimi" collection).

Overview

The purpose of a block overview is to provide enough information to enable curriculum-design teams and others to find out quickly what is meant by a block title and to decide whether to examine the candidate block. In a brief paragraph, perhaps slightly larger than the college-catalog course descriptions shown in Chapter 3, the overview of a block (as the template indicates), should provide these six pieces of information: who the block is intended to serve; what the subject-matter focus is; how instruction is organized; what the time requirements are; prerequisites, if any; and the block's rationale.

Intended Students. A curriculum block may be designed for all students as part of a core curriculum or only for certain categories of students. Examples of special categories include blocks for various specializations (trigonometry, computer programming, French, music composition, etc.), advanced placement, and remediation. Typically, blocks are identified as being intended for use at a particular grade range such as K-2, 3-5, 6-8, or 9-12. Such usage does not preclude a block being used at a level other than the one specified, but it does suggest exercising caution when not following the developer's recommendation. In principle, a block could even be designated as K-12, if it were designed to be used in any or all grades.

Subject Area. The intent here is to have the overview signal the block's general subject turf, such as science, mathematics, technology, arts, humanities, or vocational or combinations of those such as science/mathematics/technology, arts/technology, or science/history. (A more complete description of specific content appears below under the heading Content.) Further turf designation here could include traditional discipline—for example, chemistry, trigonometry, architectural drawing, music composition, or U.S. history—or broad thematic areas such as environmental studies or ethics in science and medicine.

Separate subjects, traditional scheduling of two semesters per year.

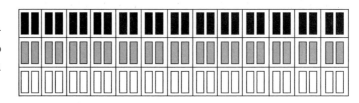

Mixture of separate and combined subjects, highly flexible scheduling.

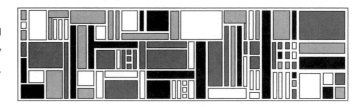

A K-12 curriculum can include uniform or varied time blocks and integrated or single discipline blocks.

Format. Instruction can be organized in several ways. The most traditional way is to organize instruction as a "course." Courses have evolved to include lectures, demonstrations, class discussion, homework, tests, and (depending on the subject matter) practical work such as laboratory, studio, or shop. Some of the possible formats are lecture/discussion course (which would typically include demonstrations), laboratory course (which adds 25 percent or more time in individual or small-group investigations in addition to lecture/discussion/demonstration) and shop course (design and construction). Other formats include course-independent design projects, seminars, independent study (individual and group), and peer teaching, which are believed to often be more effective than the traditional course format for certain kinds of content and purposes.

Time Frame. First, the overview needs to identify the total time requirement of a block—whether it is a quarter of a school year, a semester, a year, or several years, and also whether the duration is continuous or divided (for example, one semester a year for three years in a row). Then the overview should specify how the time will be divided, to include meeting frequency and duration. For some blocks, only the total time in hours may be needed, permitting great flexibility in clock and calendar scheduling. For example, a one-quarter course that is typically scheduled for an hour a day for 10 weeks would total 50 hours, which conceivably could be scheduled as densely as five hours a day for two weeks or strung out as a half-hour a week for two years.

Although the overview may specify an optimal duration for the block, it is also important to estimate the permissible range of configurations in time—that is, a range outside of which the block would have to be substantially modified or given up. Because there are human limits to dealing with complexity and change, curriculum designers must consider what balance to strike between possible benefits of curriculum flexibility and the tolerance of local teachers and students for varying clock and calendar schedules.

Prerequisites. Curriculum designers need to know what conditions, if any, students should have met before being admitted to the block in question. Prerequisites may be stated as previous experiences (other blocks to have been completed satisfactorily or, better yet, as knowledge or skills to have been acquired). In the case of blocks in the service of science literacy, *Benchmarks for Science Literacy* can serve as a convenient guide and reference.

Rationale. A curriculum block should be more than a collection of miscellaneous topics and units. It is a program of study that makes conceptual sense, to students as well as to teachers. The learning goals at which a block aims provide one kind of rationale, but students often need a more immediate sense of purpose to motivate and focus their study. For example, block rationales for students could be to explain or predict certain kinds of phenomena, design products, characterize different perspectives on societal issues. Multiple rationales could be relevant to different parts of a single block (such as understanding a given phenomenon so as to be able to consider possible results of a proposed social policy), but it is desirable to have an overall orienting rationale for the block as well. (More suggestions for block rationales appear toward the end of this chapter under the heading Ideas for Curriculum Blocks.)

"Provide a sense of purpose" is one of the criteria for block evaluation, described in CHAPTER 6: BUILDING PROFESSIONAL CAPABILITY.

Content

The content of a curriculum block can be described by the learning goals targeted, main topics treated, typical student activities, teacher and student options, and links to other subject matter. Each of these categories represents an important feature of the block that curriculum designers need to take into account. From the Project 2061 point of view, the first of those listed here, learning goals, is by far the most important, even if the next two on the list—topics and activities—seem to currently get more attention from both developers and teachers.

Stated Goals. The purpose of a curriculum block is to foster learning, and the block description should reflect what specific learning the block developer had in mind. That learning expectation must be explicitly stated in the overview to provide a basis for adopting or rejecting a block. If the block claims to address many of what the designers have identified as key learning goals, then its other properties can be examined. If not, it should be quickly eliminated from further consideration.

How specific should these goals statements be? The broader and fewer the learning-goal statements are, the easier it is for developers to claim a block in some way serves them. If a block's goal is as general as "learn about cells," for example, then anything at all about cells would do, for it is not clear what it is about cells that students should understand. Very specific goal statements, on the other hand, offer more guidance—but are likely to be more numerous and require closer attention (and be more constraining for developers). A compromise on specificity is to have the stated goals section indicate block goals by benchmark "family"—say, the section headings in *Benchmarks* or the standards headings in *National Science Education Standards* (*NSES*). That way the block's goals can be examined at two levels: first to choose candidates for a goal area and then to make final decisions using the more specific benchmarks themselves. For example, the description of a block suitable for grades 5–8 could indicate, at one level, that it deals with aspects of heredity, cells, human development, uncertainty, and models (all *Benchmarks* sections), and then at a more detailed level, list which of the benchmarks in each of those *Benchmarks* sections it actually targets. (The *NSES* parallel for this procedure would be to list the particular content standards the block includes and the "fundamental concepts" each of the standards targets.) Determining whether blocks actually do serve benchmarks well is taken up below under the heading Credibility.

All this is still at the level of claims. It should be kept in mind that science-literacy goals are not the only kinds of learning goals that a block may target. Effective and safe work habits or good citizenship, for example, can be important goals. But if they are going to be included in the block description, they should be specified as clearly as other learning goals (for example, which particular work habits or aspects of citizenship the developer has in mind).

Main Topics. Blocks can also be described in terms of what topics students will study. Students studying a particular topic may not be aware (at least initially) of what the underlying learning goals are. A topic is whatever a block seems superficially to be about—chemical equations, whales, gardens, pollution, telescopes, lotteries, epidemics, space travel. There is an important distinction to be made between Main Topics and Rationale. A block's *main topics* may be thought of as the answer that students would be most likely to give when asked *what* they are studying in a school subject; a block's *rationale* would be their likely answer when asked *why* they are studying that topic in particular. And both topic and rationale need to be distinguished from specific learning goals (either the developer's stated goals or benchmarks identified through a careful analysis), which may or may not be shared with students. A good example of a main topic is acid rain. Although acid rain is nowhere mentioned in *Benchmarks* (or in *Science for All Americans*), it would be possible for a block ostensibly about acid rain to target a variety of specific benchmarks in measurement, technological and social trade-offs, water cycle, statistics, mathematical models, and so on. Moreover, several blocks called "Acid Rain" could have different primary rationales—for example, understanding chemistry in the atmosphere or following public debate on pollution or tracing the history of industrialization.

The distinction between different meanings of "topics" is discussed further in Chapter 7: UNBURDENING THE CURRICULUM.

Activities. In addition to describing the topics to be taught and learning goals targeted, the block description should include information on how a block is characterized by what students typically *do*. Hands-on work, experiments, field trips, group work, individual study, simple projects, long-term projects, or looking up information are all possible activities. The block description should not recite all the detailed learning activities themselves, but limit itself to describing kinds of experiences and sketching typical examples of them.

Options. It is possible, even desirable, for blocks to range from those in which everything is spelled out in great detail, leaving few instructional options for the teacher, to those in which the instructions are quite general, presenting teachers with freedom to choose or invent. The same range pertains to student options; some blocks can have all students doing the same thing, others can give them choices. Most blocks are likely to be somewhere between these two extremes. Sometimes options may merely point to other resources, and sometimes all the resources for an option may be provided within the block itself. Option information in the block description enables taking local circumstances and preferences into account.

Connections. It is helpful in curriculum design to know whether a block under consideration has useful conceptual or operational relationships to other blocks. The block description should list complementary blocks that target different but related benchmarks, blocks that contain reinforcing benchmarks, blocks that include precursor benchmarks, and follow-on blocks whose benchmarks extend those in the block being described. Teachers wanting to cross subject-matter boundaries would like to know that a block has been designed to connect gracefully to blocks in other subjects—for instance, a physics block that features Galileo connects to a history block that focuses on the Renaissance. Blocks can also be related by common contextual components, quite aside from specific content—the same measuring apparatus, the same analysis technique, the same community site, or the same historical period.

Operation

To operate effectively as their developers intended, all blocks require human and material resources, time and money, means for assessing student progress, and professional preparation. Block descriptions should provide the following information:

Human Resources. Above and beyond the personnel it takes to run a school, each block in a curriculum has its own personnel requirements, which are the key to its success. The staff of a course or subject is often one teacher, but not always. A block may call for a team of teachers or for a teacher plus one or more teacher assistants. A block may also have been designed to be taught by a teacher plus a student (or team of students) or by an adjunct teacher. To increase efficiency or improve instruction, some blocks are intended to be lead by project directors, seminar leaders, or independent-study coaches. In all of these cases, of course, relevant experience and skills may be more important to consider than job titles.

Material Resources. A block description should specify the print, electronic, audiovisual, physical, and living materials that are essential, as well as those that are recommend-

A vivid display of links appears in the dozens of growth-of-understanding maps in Project 2061's Atlas of Science Literacy.

EXCERPTS FROM AN ACTUAL BLOCK DESCRIPTION

Title *Chemistry That Applies*

Overview
 Intended Students: grades 8, 9, or 10
 Subject Area: physical science
 Format: course
 Time Frame: 6-8 weeks, 5 meetings/week, 1 period/meeting
 Prerequisites: Not specified (but student misconceptions are noted and addressed)
 Rationale: To learn to develop empirical tests of hypotheses; to design and conduct scientific investigations....

Content
 Stated Goals: Unit's "Philosophy and Rationale" section lists these goals— matter is conserved in all chemical reactions; all matter is composed of atoms that join together to form molecules; new substances form when the atoms of the reactants come apart and reassemble in new arrangements....
 Main Topics: Describing Chemical Reactions/Weight Changes in Chemical Reactions/Molecules and Atoms/Energy and "Boosters"
 Activities: Predict weight changes and design experiments to test them; write equations to represent chemical reactions; use marshmallows and toothpicks to build models of molecules....

Operation
 Assessment: "Think and Write" sections check student understanding early in each lesson, with answers. End of unit exam is provided, with answers.

Credibility
 Benchmark Analysis: Rated satisfactory in a 1999 evaluation by an independent review team as part of Project 2061's middle grades science textbook evaluation. Includes content relevant to the following benchmarks: 4D 6-8#1 atomic/molecular structure of matter; 4D 6-8 #7 conservation of matter in shuffling atoms; 4E 6-8 #7 chemical energy.... Provides oustanding instructional support for benchmark 4D6-8#7....
 Development: Michigan Science Education Resources Project, Michigan Department of Education

Parts of a description of an actual (though rather short) block.

ed but optional. If special facilities or events are essential—not just desirable—the block description should alert school districts to that need. A course built around field studies, for instance, ought not to be installed in the curriculum at all if suitable nearby sites—or safe transportation to more distant sites—are not available. The description should make clear that books alone would not be adequate. The same can be said for technology blocks that require a shop with power tools, science blocks calling for laboratory investigation, blocks that rely on student access to computers. An actual block package should contain either all of the needed materials or all of the information needed to enable schools that wish to adopt it to acquire the required materials.

Assessment. How student progress will be monitored is an issue that is almost as important as quality of instruction in the view of most teachers, students, parents, and school administrators. Does the block include assessment materials linked to the targeted learning goals? What assessment techniques are integral to guiding day-to-day instruction and what are appropriate to longer-term summary judgments about student or program success? Are they objective tests? Essays and reports? Interviews? Portfolios? Is guidance provided on scoring and grading? Do the assessment materials provide suggestions on how to use the assessment results to revise the block?

Teacher Preparation. A block description should make clear what special knowledge or skills a teacher (or whoever is to teach the block) should have beyond what is ordinarily expected. A science course, for instance, may call for the teacher to be familiar with certain episodes in the history of science, knowledge that even well-prepared science majors may lack. A block may demand skill in supervising student research projects. To this end, a block may include professional development materials (print, audiovisual, computer, on-line) or it may cite readily accessible books, articles, films, computer programs, or other resources.

Cost. Costs vary from course to course, although it is not clear how often costs are taken into account in making curriculum choices. If block descriptions give fair estimates of financial and time costs, then design teams can at least make rough comparisons of the cost-effectiveness of contending blocks. Although benefit-cost analysis is not highly developed in education, even rough approximations are useful, and the process of conducting such analyses can be counted on to at least lead to a more penetrating consideration of curriculum decisions than if costs are ignored.

Credibility

The learning goals intended (or at least claimed) by block developers will already have been cited under "Content" above. But materials do not always correspond well with the goals claimed for them. How confident can one be that the claims will actually be achieved? Scientific evidence of student learning is rarely obtainable, but confidence in the credibility of the learning claims made for a curriculum block can be strengthened considerably by rigorous benchmark analysis. In the absence of such analysis, some suggestions of effectiveness may be provided by uncontrolled empirical studies of student learning, published reviews in professional journals, and the testimony of users. If the developer of the block has demonstrated high credibility in other blocks, that too may provide a clue.

Empirical Evaluation. Ideally, the evaluation of curriculum blocks ought to be based on concrete evidence of what students who experience the instruction really do or do not learn. Empirical research studies are difficult to conduct and very costly in the bargain, but sometimes they are carried out and the results are made available in professional journals and institution reports. But even then there are many cautions, including bias in the selection of comparison groups. Oftentimes, for example, research on instruction is conducted on particularly able and interested teachers and students, which leaves open the possibility that the materials and methods may be significantly less effective in less special circumstances. Therefore, reports on research should describe who the trial teachers were (from development participants and volunteers at one extreme of specialness to randomly selected samples at the other) and how special the students were in previous academic achievement and motivation. Also important is the nature of the actual assessment tasks and how long after instruction they were administered. This reflection on evidence can be seen as a step toward putting judgments about curriculum on a more scientific footing.

Benchmark Analysis. Short of scientific studies of what students actually have learned, but more credible than general opinions or published reviews, are systematic analyses of curriculum-block materials and activities to estimate what students would be *expected* to learn from them. A careful analysis based on principles of how students learn is not an easy or rapid undertaking. Even experts have a tendency to make general, impressionistic judgments about whether the content seems to feature benchmarks and use sound instructional approaches—judgments that they often reverse when they are required to cite specific evidence in a block's materials and activities (or notes to teachers) that those *particular* benchmarks would likely be achieved through *particular* activities.

PROJECT 2061 CURRICULUM-MATERIALS EVALUATION PROCEDURE

Step 1: Identify benchmarks that appear to be covered by the curriculum material.

Step 2: Clarify the benchmarks' meaning.

Step 3: Reconsider how specifically the material targets the benchmarks.

Step 4: Estimate how effective the instruction would be.

Step 5: Recommend improvements.

Project 2061 has organized expert analysis by educators experienced in the procedure to make possible a sort of "consumer report" on popular curriculum materials in mathematics and natural science. Familiarity with the analysis will nonetheless be necessary for educators, parents, and community leaders to make the most of the reports. For more information, see the Project 2061 web site at www.project2061.org.

A valid and reliable analysis requires just that: identifying evidence in instructional materials and teacher guides that the block would be effective in helping students to achieve specific benchmarks. In the future, we can expect that teams composed of teachers, curriculum specialists, and university researchers—all of whom have undergone extensive training in such benchmark-by-benchmark analysis of materials—will be able to examine blocks in detail and report their findings in a concise form that practitioners can use in making block-selection decisions. As such reports become available, they should become essential parts of block descriptions.

Reviews. In the absence of empirical studies or systematic analysis, or as a supplement to them, personal-opinion reviews can be of some value. Some teachers and scientists are exceptionally good at critically reviewing learning materials. For decades, the journal *Science Books & Films* (published by AAAS) has successfully called on such aficionados to write reviews of trade books and other science materials that have been prepared for children and the general public. The *Science Books & Films* reviewers comment on materials' scientific accuracy, readability, interest, and suitable audiences. Similar reviews could be published for textbooks and curriculum blocks, with additional commentary on ease of use and appeal to teachers. As reviewers become more familiar with the criteria for effectiveness used in benchmark analysis, they can tune their work more closely to estimating what students would actually learn. Then, as the education profession matures, its journals will certainly devote substantial space to expert reviews of curriculum blocks. As such reviews become available, block descriptions will be able to cite them.

Users. Curriculum designers considering a particular block will be interested in who has already judged the block favorably enough to adopt it—and what they have to say about how it has worked out for them. A block description ought to indicate whether the block developer maintains and makes available an up-to-date list of users in different regions of the country. Such a listing would make it convenient for a curriculum-design team to communicate directly with users in school districts similar to its own. Interviewing teachers who have actually taught the block can bring to light useful information, both positive and negative, not found in any other way.

Development. It matters how and by whom a block was created. The block description should indicate how teachers and scientists were involved and with what kinds of students the block material was tested in classroom settings. The description should also refer to published research on teaching and learning that support the block's instructional strategy. The more that scientists and classroom teachers are involved in the block's development, the more extensive and representative the field testing, and the more rounds of revision in response to that testing, the greater the confidence that can be placed in the block being able to deliver the learning outcomes that are claimed. Serious attention to cognitive research, when explicitly and specifically claimed (or done) by developers, is also an important source of credibility.

In the short run, the task of describing blocks in these template terms will likely fall on specialists who set up the database. In the long run, developers of blocks, perceiving consumer interest in using the database, may do much of the description work themselves— especially once they have begun to build in the desirable properties that the template implies.

Inevitably, a large curriculum-block database—perhaps a single, national one—would include blocks of differing quality. The objective is to include blocks in the database on the basis of the completeness of a block's description, not on a definitive conclusion about its quality. Local school districts, not distant agencies or organizations, will decide which blocks merit adoption.

Such explicit research on student learning is summarized in CHAPTER 15: THE RESEARCH BASE of *Benchmarks for Science Literacy* and in *Resources for Science Literacy: Professional Development.*

Hagar the Horrible by Dik Browne

WHERE WILL CURRICULUM BLOCKS COME FROM?

The design-by-assembly approach calls for a large national pool of curriculum blocks, that are well described in an accessible database. But where will the blocks come from? Who will actually create them?

Developing curriculum blocks is a job for teams of specialists: scientists, mathematicians, historians, engineers, statisticians, psychologists, educational researchers, artists, literary critics, and other scholars and practitioners. It also takes graphic artists, writers, activities inventors, filmmakers, computer-program writers, test makers, and others. Among the most important specialists to have on the team are classroom teachers.

This book, as already explained, merely supposes that such an inventory of blocks and database of descriptions will become available in the early years of the 21st century and does not offer any direct advice on how to develop them. But here are some thoughts on how the actual development of curriculum blocks could come about.

Sources of Curriculum Blocks

Curriculum blocks can be designed by modifying some of the creditable courses and units currently in the curriculum. American elementary and secondary education is blessed with a large number of highly creative individuals—classroom teachers, professors, grantees of public agencies and private foundations, and editors and writers for publishing houses. Some of them have succeeded in developing courses or smaller modules that, with some work and with complete and accurate descriptions, could become curriculum blocks of the highest quality. Perhaps the most crucial aspect of this task would be tying these existing courses—operationally, not rhetorically—to specific learning goals such as those in *Benchmarks*.

Another potential source of curriculum-block material consists of some of the courses developed during the era of curriculum reform a generation ago. For one reason or another, many of them have gone out of use or have left only a few traces in other materials. With a modest investment of time and money, some could be recast to take advantage of the greater flexibility now possible in format and time, the advances in research knowledge about how students understand and learn, and the availability of more powerful technologies. Indeed, some of the nonprofit organizations under whose auspices the courses were originally produced continue to be world leaders in the development of innovative instructional materials. Given suitable financial support and incentive, those organizations could expeditiously and creatively undertake this refurbishment.

But in the long run, a large number of altogether new blocks will have to be created to provide more choices for curriculum design tailored to a spectrum of populations, resources, and viewpoints. They can be developed starting with interesting ideas on instruction, with a selection of specific learning goals, or with readily available materials, particularly trade books.

As a way to stimulate thinking about curriculum blocks, Designs on Disk contains a database for creating block descriptions.

The Role of Teachers in Developing Curriculum Blocks

It is sometimes thought that classroom teachers create the courses they teach and that the teachers in a school create the school's curriculum. In truth, very few teachers design courses *de novo*. Instead, they shape a given course—5th-grade arithmetic, say, or 9th-grade earth science—to fit their prevailing circumstances (students, resources, policies) and to reflect their professional judgment. This invaluable contribution to learning acknowledges the fact that such shaping is almost always needed, which is one reason that good teachers will always be essential. Teachers also are a key source of ideas on what is needed and what students are likely to respond to well. Some teachers, of course, become expert contributors to development teams. And many teachers are needed to carry out the field testing of prototypes and provide systematic feedback on the blocks once they are in use. But to expect all teachers to develop their own curriculum blocks is unreasonable. Few have time or resources or experience for such an undertaking and are better occupied in fine-tuning commercially available blocks for their own students.

However, it is neither possible nor desirable to have all new curriculum blocks developed by teams or organizations. There is the obvious limitation of resources, for those groups can do development work only if they are able to secure adequate financial backing. The availability of such support varies greatly over time and often does not last long enough to sustain steady curriculum development. The commercial sector, on the other hand, has the resources to provide the needed research and development capital and is able to assemble the creative talent needed to do the job. Many publishers, television stations, filmmakers, and software developers produce innovative materials and have a strong market incentive to meet new curriculum demands. But they too must include teachers on their development teams.

Realistically though, how can we expect teachers to take part in such research and development (R&D) when their day is already taken up by teaching, preparation, and dozens of other duties associated with schooling? Although most teachers do some R&D in their own classrooms, there is no tradition in the schools—comparable, say, to that found in research universities—to justify such work, and no provisions for supporting it. And even if there were, few teachers have the training or experience to conduct R&D at a high professional level.

To some degree, the participation of teachers in creating curriculum blocks could be arranged by allowing them to become members of funded development projects based in universities or professional societies. That happened in the 1960s with great success. Not only were the products of those curriculum projects much better than they would have been without teacher input, but the nation ended up with a cadre of trained curriculum specialists. The process should be reinstituted, this time with the understanding that the teachers who participate will return to teaching in specialized institutions in which they can continue to engage in R & D. And they should be sought out by those publishers seriously intent upon investing in state-of-the-art product development.

For financial and historical reasons, few of the 15,000 school districts in the United States are able to undertake R&D activities on a sustained basis. Hence, teachers with applicable experience-based preparation have nowhere to go to utilize their new knowledge and skills. The notion of the "R&D school district" holds that out of all of the school districts in the nation, we need to transform only a relatively small number—perhaps a few dozen—into formal R&D school districts—that is, school districts that, as a matter of policy, would participate in R&D activities of one kind or another on a continuing basis. (All school districts, however, will continue to do local R&D in adopting national products for their own contexts and students.)

Some of the varied arrangements known as "professional-development schools" already include components of R&D. Also, there is collaboration on R&D between regional education laboratories and schools. Some R&D school districts could be just such hybrids, analogous to "research universities" or "teaching hospitals," both of which typically have parallel and mutually supporting roles of education, research, and development.

Some university faculty members do conduct research studies in schools, and specially funded development projects do test their own materials and courses in schools. But in-school faculty research is not the equivalent of clinical research conducted in research hospitals (most hospitals are not research institutions), and in-school testing is rarely very systematic in design. And just as the clinical investigators in a research hospital can call on the help of the basic researchers in the university with which it is associated, and are able to scour the research literature for knowledge that can be applied to medical practice (which most practicing physicians are not prepared to do), so too would "clinical researchers" in R&D school districts. The main point here is simply that it will take institutional change to enable teachers to participate effectively in developing the curriculum components of the future.

The experience of Project 2061 strongly suggests that some such arrangements will pay off. Six Project 2061 school-district centers were established in 1989 to provide the project with colleagues who would contribute to its work in creative ways. Each center's team members had the freedom to participate in R&D activities that were not necessarily aimed at improving instruction in their districts. In other words, they were not set up as traditional demonstration sites but as district-based R&D centers with university ties and a national focus.

The Project 2061 school-district team members have had a major hand in developing *Benchmarks for Science Literacy, Benchmarks on Disk, Designs for Science Literacy, Designs on Disk, Resources for Science Literacy: Professional Development, Resources for Science Literacy: Curriculum Materials Evaluation,* and *Blueprints for Reform* as well as Project 2061's growth-of-understanding maps. The team members have been in on the conception of these reform tools, have provided many of the key ideas in them and helped to clarify them, and have undertaken a critical appraisal of all of the draft documents. They have not been merely helpful, they have been essential players in what is at heart an R&D effort. Given the opportunity and the resources, teachers in other districts could be equally productive, in time ameliorating the shortage of teachers prepared to participate in such efforts.

Ideas for Curriculum Blocks

As the curriculum-block idea gradually took shape, along with clarification of the curriculum-design concept and emergence of the notion of curriculum design described in the previous chapter, Project 2061 began to accumulate suggestions for blocks *to be created.* Ideas came from Project 2061 staff, the school-district team members, and consultants. In some cases, contributions were complete enough to be considered specifications for block development, though most were less complete than that.

Below are some categories of ideas to illustrate the variety of curriculum blocks that could well be part of the national inventory. They are not offered as a set of definitive or mutually exclusive categories, but as emphases that would allow many mixes to be worked out in detail.

Applications blocks emphasizing the uses of science, mathematics, or technology. Examples: Chemistry and Society, Public Opinion Polling, and Science and Crime Detection.

Case-study blocks in which the content is organized around one or more case studies that focus on historical episodes, social issues, or technological problems. Examples: Darwin's Finches, The Chemical Revolution, and Brecht's Galileo.

Design blocks organized around design challenges for students to respond to individually or in groups. Examples: Energy Conservation, Measuring Time, Remote Controls.

Cross-cutting blocks that link science, mathematics, or technology to other domains. Examples: Architecture, Dinosaurs and Dragons, the Panama Canal, Evidence in Law and Science.

Discipline blocks that present some important aspects of the content, methods, and conceptual structure of a discipline. Examples: Anthropology, Statistics and Probability, Biochemistry.

Explanation blocks that are designed to help students understand phenomena, objects, and systems. Examples: Fire, Growth and Decay, Science and Technology Underground, and Plagues.

Exploration blocks that examine a place or time from the perspective of science and technology. Examples: Science and Technology in Ancient Egypt, the Lewis and Clark Expedition, Science Underwater, and Science in Space.

Inquiry blocks that engage students in designing and carrying out scientific investigations to foster an understanding of how science goes about its work. Examples: Objects in Motion, Neighborhood Insect Species, and Traffic Patterns.

Issue blocks, usually seminars, that engage students in an examination of issues involving science and technology. Examples: Ethics of Experimentation, Genetic Engineering, and Populations.

Theme blocks based on broad concepts—such as those in CHAPTER 11: COMMON THEMES in *Science for All Americans* and *Benchmarks for Science Literacy*—that cut across science, mathematics, and technology. Examples: Feedback in Biological and Social Systems, the Size of Things, and Evidence.

> Any of these "types" of curriculum blocks might be provided in different forms—a mailable package, an Internet site, or access to the resources of a particular institution.

When curriculum blocks are thoroughly and accurately described as to specific goals targeted, operational requirements, and evidence of success, school-district curriculum-design teams will have the information they need—now almost completely lacking—to select a set of blocks that will meet local and state requirements. A district will construct its curriculum from a relatively few of the blocks, but different districts could end up with different arrays of blocks and still meet national, state, and local standards.

LOOKING AHEAD

A national computer database of complete curriculum designs will eventually be available as an alternative starting place for curriculum designers. Using computers, designers will be able to draw on libraries of specific learning goals, fully described curriculum blocks, curriculum-design concepts (many of them elaborated well beyond the concise examples we have given), and complete curriculum designs. To top that all off, they can expect to have access to information about school districts that have actually implemented specific curriculum designs and how those designs are turning out. What is learned from the implementation of designs can then be used to improve the designs for use by others and to enhance our knowledge of curriculum more generally.

How will the new school curricula of the future be created and implemented? The next chapter presents the still-imaginary stories of how that will happen in three different school districts of the 21st century.

Thomas Arledge, *Mailboxes*, 1999

CHAPTER 5
HOW IT COULD BE: THREE STORIES

The purpose of these stories is to suggest how curriculum reform could proceed using the ideas laid out in the preceding chapters. Written from the perspective of a time a decade or so in the future, they tell how three quite different school districts went about creating new designs for their K-12 curricula. The Palladio Unified, Edmond Halley, and Lewis & Clark Regional school districts don't actually exist, except in our imagination, so we can purport to know everything about them and exactly what transpired. The stories in this chapter are most directly about the science, mathematics, and technology components of the curriculum, but much of the account is relevant to other subjects and to the entire curriculum as well.

Never mind that these stories are made up. The intent here is to provoke fresh thinking about how we can get beyond marginal change in curriculum, not to pre-scribe a sure cure for current curriculum ills. Although in each case the reform process is presented as a series of steps, in the real world of schools, things happen in parallel and in loops. Stepwise presentation merely makes it possible to consider and describe the components of curriculum reform. As in architectural design, curriculum design draws as much on the spirit of art as on the spirit of engineering, and hence it inevitably involves compromise between form and function, between respect for the past and an eye for the future, between intellectual integrity and political reality.

Knowledgeable readers will recognize that these tales make no effort to deal with any number of serious impediments to the realization of reform in K-12 education. The tales focus narrowly on the process of K-12 curriculum design writ large—not on turning a curriculum design into an actual curriculum and not on operating a curricu-lum once it is place. In fact, of course, little of that can possibly happen unless the conditions listed in Chapter 3's Setting the Stage section are realized: consensus on the need for reform, well-developed and evaluated curriculum blocks, better prepared teachers and administrators, supportive policies, and adequate resources.

These three stories simply assume that those matters have been dealt with satisfactorily. Reality may turn out otherwise. In any case, it is not the purpose of this volume or these stories to propose solutions to the many problems that beset American K-12 education. Rather, it is to think about how it may be possible in the future to actually go about the business of designing the entire K-12 curriculum as a whole in terms of specified learning goals. The approaches described here do not constitute Project 2061 endorsements of best practices, or even portray the full range of possibilities. They do raise issues that any real design enterprise will have to address.

FOREWORD

We are in the future looking back. In undertaking to create new curriculum designs, the three school districts had begun a decade earlier with one crucial advantage in common: sustained purposefulness. The district leaders seemed to know that major curriculum reform couldn't simply be mandated from the top down, the bottom up, or the outside in. They realized that first there had to be a widely shared desire for change among the administrators, teachers, parents, and citizens who ultimately would have to approve of undertaking any serious design effort, participate in the design process, and then see to its implementation. They knew that impeding change is vastly easier than effecting it. So the districts were willing to invest, in their various ways, the time necessary to secure the involvement and support of key individuals and groups.

But desire for curriculum change, no matter how widespread, will not itself ensure the creation of effective designs. Indeed, designs cobbled together in a hurry by unprepared designers are very likely to result in curricula that are worse than the ones they replace. So the districts also took the time necessary to build a capability for curriculum design and change.

Fortunately, building capability and building support go hand in hand. Rather than start out by talking about the need for grand changes in the curriculum and proposing radically new possibilities, each of the three districts first examined its entire curriculum in light of accepted content and structural standards, identified its strengths and weaknesses, and reported its findings regularly to the entire faculty, school board, and community. Each such study took several years because it dug into the details and because the participants had to learn how to carry out sophisticated analyses. As the study reports appeared, more and more interested parties were drawn into thoughtful and informed discussions of what the curricula were actually like and what about them most needed attention.

How the districts went about that work varied according to their size, circumstances, and traditions, but all three drew heavily on their state frameworks and nationally accepted goals statements. In the context of science literacy, these included Project 2061's *Science for All Americans* and *Benchmarks for Science Literacy;* the National Research Council's *National Science Education Standards;* the content standards produced by the National Council of Teachers of Mathematics, International Technology Education Association, and the National Council for the Social Studies; and similar documents in other subject areas. Moreover, as time and money permitted, representatives attended appropriate conferences and visited other districts that were said to have outstanding curricula or that were experimenting with novel curriculum components. During those same years, professional curriculum-materials developers invented a large number of diverse curriculum blocks that were field tested, revised, and fully analyzed in a national computerized database.

Even as the curriculum studies proceeded, teachers in each of the three districts were making small improvements in parts of their district's curriculum. They tried out new teaching techniques and time arrangements, sampled alternative teaching materials, introduced some new content and, although it was harder to do, eliminated some topics, even some previously untouchable ones. With some material eliminated, they worked on how to use the extra time to teach better some important ideas that students had not learned well before. Teachers kept records on what innovations they tested and shared their findings with one another. In short, the districts followed many of the suggestions found in Part III of this book on how to get started toward long-term curriculum reform.

Because of their participation in these activities carried out over an extended time, a large proportion of the teachers and administrators in the districts became knowledgeable about curriculum-design issues, possibilities, and limitations. Further, a strong consensus developed on the need for a major curriculum transformation and a strong sense of what direction that transformation ought to take. The faculty deepened its understanding of curriculum architecture and became skilled in the use of computer-based tools for curriculum design and resource analysis. During those years, newly hired teachers were assigned to serve as apprentice members of the various curriculum committees and study groups, so that their continuing professional development was linked from the outset to reform activities.

The decade of focused preparation in our three districts was carried out in the spirit of exploration and learning, and it engaged students, parents, and concerned citizens in all of their cultural, ethnic, economic, and occupational diversity. In none of the districts was total harmony realized among the educators or within the community, but enough agreement eventually emerged—after some give and take on one aspect or another—to warrant moving from piecemeal improvements and cautious experimentation to systematic, comprehensive curriculum redesign. For more on what happened, we turn to each of our three districts: Palladio Unified, Edmund Halley, and Lewis & Clark Regional.

PALLADIO UNIFIED SCHOOL DISTRICT

Palladio Unified School District is located in a very large, heavily populated urban area. Its student population has changed dramatically over the decades, largely reflecting the waves of migration that have passed through the city, and its financial health has mirrored the ups and downs of the region's industrial base. Currently, African-American, Hispanic (mostly Central American and Caribbean), and Asian-American (mostly Southeast Asian) students make up most of the Palladio student body, although the overall proportion of those minorities in the city itself is considerably less. More than a third of the students come from working-poor and under-employed families. In spite of federal aid and state equalization funds, the school district is unable to secure the financial resources it feels it needs to put it on a par with the school systems in the surrounding suburbs—or, more critically, to reach the ambitious goals it has set for itself.

Nevertheless, this is not a school district in disarray. In its century and a half, it has faced severe challenges time and again, enjoyed its share of successes, survived its failures, and believes that such is its fate. Palladio Unified exists in a very political city that never hesitates to criticize its institutions, a city that is often at odds with the state and the courts over school matters. Yet surveys show that a substantial majority of parents and citizens believe the schools are making headway in the face of the city's current challenges. One reason may be that because they live in a city of neighborhoods (some of which were formerly villages bordering the city proper that were incorporated into the city against opposition that has never entirely subsided), parents and citizens focus more on *their* particular schools than on "the system," and they do not think that criticism (or even praise) of the school system as a whole has much to do with them.

Because of that attitude, and because the district's rather rapid turnover of superintendents and school-board members has made it difficult to create and sustain a strong central bureaucracy in any case, Palladio has never become strongly centralized.

Thus, the district's name not withstanding, creating a single districtwide curriculum has never been in the cards for Palladio Unified. With regard to this curriculum-design undertaking, everyone understood that the role of the central administration and board was to set directions, provide incentives for action, distribute resources fairly, monitor progress, and take care of legal and financial oversight.

Settling on a Process

In a series of joint meetings in various sections of the city, the Palladio school board suggested to teachers, principals, and parents that after nearly a decade of making worthwhile but piecemeal improvements in the curriculum, a more thorough and coherent curriculum was in order. As increasingly close attention had been paid to specific learning goals, it had become more and more evident that students were not learning well. Moreover, the faculty now had the ability to design a curriculum. This board declaration was neither surprising nor unwelcome. It was followed by a series of hearings with representatives of the teachers' and principals' organizations, the city's political and business leaders, and the general public. The aim was to ensure enough political support to effectively carry out a major curriculum-design effort.

The district plan of action that emerged had these main features:

- Although there was strong support among teachers, principals, and parents for making curriculum changes on a larger scale than before, relatively few of them believed it was desirable to create an entirely new curriculum. Thus, the leaders provided for two parallel lines of action: (1) to carry out a major redesign of the existing curriculum *school by school* along traditional lines for one set of schools and (2) to create an entirely new, highly innovative K-12 curriculum for another set of schools. Each school could decide for itself which line to participate in. Teachers were given the option of transferring to other schools, so they could participate in the design undertaking of interest to them.

- Each school would have a budget to underwrite its curriculum-design work and would be free to go about that work however it wished—as long as it was making reasonable headway, as judged by an advisory panel established by the school superintendent. However, since much could be gained from collaborating with others, some additional funds would be made available to encourage schools to form voluntary groups to share in the task.

- To ensure that this design freedom would not result in ineffective curricula, whether traditional or new, the school superintendent would submit for board approval a single set of essential learning goals that had to be met at various assessment checkpoints. Acceptable designs would have to demonstrate that they would enable students to meet district standards of accomplishment. Moreover, the common checkpoints would provide some assurance that students from several lower schools could be successful in the higher grade school they fed into. (It was recognized that students who changed schools at years between the checkpoints, however, could encounter some mismatch.)

Conceptualizing a Design

It followed from the decision to proceed on two separate tracks that two design concepts would be needed, one for the "major redesign along traditional lines" majority and one for the "essentially new design" minority. Some individuals realized that although this dual approach could be seen as giving up on any chance of ever achieving district unity, it could also be seen as a clever alternative strategy to the usual one of transforming an entire district from scratch. Those people also believed that if a more radical design actually turned out in practice to outperform the other, then a second set of schools in the district could replicate it, and later a third set, and so on until the entire district had been transformed—all without ever requiring the reluctant majority to make the change.

Traditional curriculum. As work got under way, it turned out that few of the elementary and middle schools wanted to work in total isolation and so groups of schools gradually formed partnerships. Some groups were composed of schools of the same grade ranges, some of schools of adjoining grade ranges. A consensus soon emerged to the effect that "traditional" meant only that the curriculum in any one elementary or middle school would be organized by the usual subjects and disciplines and in uniform time configurations, and that all students in that school would experience the same curriculum. Otherwise, school independence would prevail: external similarity, internal diversity. As this concept played out in the high schools, each school decided to remain a traditional comprehensive high school, but, for the first time in anyone's memory, each would design its entire curriculum—four years and all subjects—simultaneously and would do so without having to match the other schools.

New curriculum. Here the challenge was somewhat different. The teachers, principals, and parents agreed to create a curriculum design more or less from scratch and to approach it from the start as a K-12 undertaking. The design concept they finally settled on was a variation on the theme of diversity. Instead of the school-by-school curriculum diversity their colleagues sought, they had instructional diversity in mind. They argued that since different students generally respond differently to different ways of teaching, and different kinds of content are better taught by some instructional approaches than others, the curriculum should be designed to incorporate a variety of ways of organizing subject matter, partitioning time, and formatting instruction.

But replacing the existing curriculum with a hodgepodge was not what anyone had in mind. Unconstrained diversity would almost certainly end up as chaos and lead nowhere. So the design groups eventually adopted a core-plus-late-diversification model. The core curriculum would be designed to be completed by the end of the 10th grade, leaving students free during their last two years to enroll in one of several special schools. These schools, which might be housed together but run independently, would be two years long, but affiliated with a community college to make available three- and four-year programs. To start with, the options would include a science and engineering school, a performing-arts school, and a health-sciences school, each of which would be open to both vocationally oriented and college-bound students who had come successfully through the new ten-year curriculum.

Selecting Curriculum Blocks

The job of deciding what kinds of curriculum blocks would be eligible for selection in designing the Palladio Unified's curricula turned out to be relatively easy. Most of the schools involved found it expedient to form groups and share the burden (and risk). And the group that set out to create an entirely new curriculum necessarily had to share in specifying criteria, given the schools' commitment to unity. But what made the job so manageable was less the sharing of the labor than the nature of the decisions that had been made earlier.

Goal specification. Everyone engaged in the design process was aware of the board decisions with regard to goals, and few contested them. The board had made it clear that, although it had adopted some goals for operations, such as reducing the district's dropout rate, the paramount goals were to be learning goals—namely, what students were to be expected to know and be able to do at various stages of their schooling. For guidance on what those learning goals would be, the board had officially adopted the current state framework, which, it noted, was said to be consistent with the national standards in the various school subjects.

Like many such documents, however, the state framework was cast in terms that were too general for the purposes of curriculum design. It is not easy to derive specific goal statements from general ones, but fortunately that had already been accomplished in most of the subject areas. In natural science, for example, the CD-ROM *Resources for Science Literacy: Professional Development* (a tool well known to the Palladio teachers) contained a computer utility that identified the "fundamental understandings and skills" statements of the National Research Council and the "benchmarks" of Project 2061 that corresponded to the content recommendations found in the frameworks of each of the states that had issued them. With the greater specificity of these nationally accepted goals, the job of selecting learning goals was essentially complete. The more formidable task lay ahead—understanding the goals deeply enough to use them thoughtfully in the design process.

Constraint specification. Limits on what the Palladio Unified curriculum could be like came from near and far: the community, the state, the nation. It was recognized that such constraints might have the authority of law, such as the number of days of school, or they might be matters of local custom and finance (for example, limits on field trips). The superintendent was responsible for seeing that design teams were informed about the constraints—the number was really very small—and for checking the draft designs to make sure the constraints were dealt with properly. In the traditional-curriculum case, the main constraints involved decisions to limit the curriculum to discipline-based courses and to require uniform time arrangements. In the new-curriculum case, they involved the requirements that a variety of instructional formats be used and that the curriculum as a whole have several different alternative specializations in the last two years of high school.

Block selection rules. With the goals and constraints identified and understood, the school groups formulated two selection rules to be applied as follows:
- **Both groups**—the blocks selected had to target all of the learning goals specified for each of the chosen grade ranges—K-2, 3-5, 6-8, and 9-12.
- **The traditional-curriculum group**—all blocks had to be discipline or subject based, to be a semester or year in length, and to have the same time configuration. Grades K-8 had to contain only core blocks, but grades 9-12 could include up to 30 percent elective blocks.
- **The new-curriculum group**—starting with grades 3-5, every grade range had to include at least one seminar block and one project block. Blocks could vary in length from one quarter to three years and have any time configuration. Core

blocks would be distributed so that they would account for all of grades K-2, about 90 percent of grades 3-5, 80 percent of grades 6-8, and 70 percent of grades 9-10. Grades 11-12 (and in some cases 11-13 or 11-14) would be entirely noncore, as determined by the affiliated high school and community college. Grade by grade, the final K-10 configuration would be the same in every participating school.

The proportion of core and noncore can be represented like this:

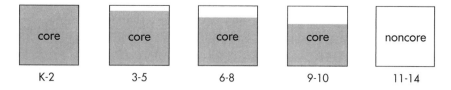

Completing the Design

Palladio Unified's curriculum design efforts had unexpected results. The group that sought school-by-school curriculum independence realized substantial unity after all, and the unity-committed group had some trouble getting there. It went like this.

The traditionalists. In principle, each school could go its own way as long as its final design could be shown to meet the district-set learning goals and honor the agreed-upon commitment to standard school disciplines and uniform time frames. But by its nature, such a loose design concept does not provide much actual guidance to the teachers, administrators, and parents at a given school when it comes to choosing blocks to build a curriculum. In this case, what happened was that each of those individuals at each school who wanted to have a hand in formulating the curriculum began browsing the curriculum-block database looking for "interesting" blocks for consideration.

Gradually the pool of such candidate blocks began to grow and "selling" began. The proponent of a particular block would try to persuade colleagues that it was better than any of its alternatives in the pool. As the elimination of candidate blocks proceeded, the configuration problem came to the fore: In each school the blocks had to fit together, but blocks selected individually, however popular, would not necessarily do so. In fact, some blocks often introduced large helpings of unwanted redundancy and left sets of learning goals unattended.

It made sense, therefore, to see what other comparable schools in the district were coming up with. Indeed, even earlier, informal exchanges had begun to take place (often to find allies for certain blocks), but the need to share ideas on the configuration problem increased the frequency of such exchanges. Knowledge of solutions worked out in one school quickly spread throughout the district and often influenced decisions made at other schools. The result of these informal exchanges was that the designs finally adopted by the schools (of similar grade range) were more alike than different, but each school could point to some differences that set it apart. Moreover, consultation between schools at different grade levels (often to join forces in support of blocks that related well to other blocks being adopted in higher or lower grade ranges) led to more K-12 continuity in the end than might have been expected. The board approved all the variants.

The avant-garde. This smaller group necessarily had to work more formally, since it was committed to K-12 unity from the start. Recall, however, that the unity sought was mostly structural—emphasis on a common core of studies that met all district-specified learning goals by the end of the 10th grade and a single pattern of instructionally diverse blocks (project and seminar blocks as well as traditional courses). That left open such matters as (1) the proportion of project and seminar blocks in each grade range, (2) how the blocks would be configured, (3) what the balance would be between discipline-based and integrated blocks, and (4) how much variety in time frames would be permitted.

A steering committee elected by the teachers identified six issues for further study before beginning to search for curriculum blocks. Teachers and administrators could participate in the study of as many issues as they wished, but by common agreement had to contribute to at least one. The six study groups, each of which had participants from every grade level, focused on the use of (1) time, (2) content variety, (3) instructional variety, (4) student differences, (5) K-12 articulation, and (6) the nature of the non-core. The purpose was to further clarify the design concept—unity with variety—by trying, as it were, to complete a floor plan before furnishing the rooms, with the understanding that subsequent decisions on furnishings might require changes in the basic plan.

Some participants attended formal sessions, others studied the topic and submitted their ideas to e-mail files that had been established for each of the study issues. Consensus was not quickly reached within each topic, and it turned out that decisions reached on one issue were not always compatible with those reached on other issues. But with time, the differences within and between study groups were reduced, and the steering committee then appointed a small subcommittee to come up with a plan that would incorporate as many ideas as possible from the study groups and yet end up with a coherent overall design.

By that time, the process had taken the better part of an academic year. Only then did actual block selection begin. A large K-12 design team representing the multiple groups worked through the summer and proposed a complete set of blocks that it believed met the criteria, fit the plan, and was interesting. Not all individuals and groups agreed, and they offered substitutions, arguing why this or that block was a better bet than one of those proposed by the design team. In two months, with an adjustment here and there, enough agreement existed to send the design to the school board for approval. Because the design was radically different from the existing curriculum and from the design adopted by the traditionalist group (whose plan was quickly approved by the board), it took the rest of the school year to secure sufficient public support to enable the board to give its official approval.

That was the easy part. Everyone understood that implementation would be difficult, especially because the district would have to accommodate two very different designs. Planning for their implementation would have to be no less thoughtful and thorough than the design effort itself. It took a summer and another school year to work out the details of the process, and then three more years to institute the changes. It took that long for many reasons, the chief one being that no new blocks were introduced without suitable teacher preparation in using them.

Interestingly, by the time the two designs were in place, enough of the teachers and administrators in the traditionalist group—and essentially all of the newly hired teachers—had opted for the avant-garde approach to justify starting another such track. That happened about every three or four years, so that within about 15 years of undertaking the design effort, the original minority avant-garde had become the new majority traditionalists in the district. What appears to have happened is reform by co-option and attrition. Instead of trying to make radical changes in the entire system at once, the district ended up changing the curriculum by gradual substitution.

EDMOND HALLEY SCHOOL DISTRICT

The Edmond Halley School District encompasses a single suburban city of moderate size and prosperity. The district maintained a generally good reputation as it grew with the community after World War II, with few scandals, high graduation rates and elite-college admissions rates, and winning basketball teams. When the city population stabilized demographically, so did the school population. The annual turnover of students, teachers, and administrators is relatively low compared to national norms. But when

the decade of reform began, the age distribution of teachers was such that many would retire before the new curriculum design would be implemented.

Halley (with many fewer schools than Palladio) sees itself as a unified entity conceptually and operationally, not as an aggregation of nearly independent schools. The hierarchy of authority is clear and generally accepted in principle, if not in every instance. Ideas, concerns, and information flow up, policies down. Committees—standing and ad hoc—are where the action is. Because committees are accorded the time and other resources they need to do their work and are usually paid attention to, teachers and citizens do not mind serving on them.

Settling on a Process

After extensive consultation with all the various stakeholders in the community, Halley's board of education selected a task force of teachers, administrators, and citizens to take the lead in creating a districtwide curriculum design, hired a curriculum-design consultant, and assigned support staff to the task force. This activity paralleled how the same board would have proceeded if it had been planning to build a new school. The task force's charter spelled out what the task force was to deliver and provided a budget, timetable, and enough procedural details to begin to work. Directives given to the task force by the school board included, in addition to oversight and monitoring provisions, these conditions:

- The design process would be made explicit and adhered to faithfully. If at any point it seemed to make sense to alter the process, changes would be negotiated with the groups involved in the process.
- Representatives of all stakeholder groups would have a part in the design process, and, although the school board had the final say by law, no participating group would have de facto control. The process would be open to public scrutiny at all times, even at the cost of slowing things down.
- Sufficient time and money would be made available in that such a politically sensitive and conceptually difficult task could not be done cheaply or in haste without great risk of failure.

Conceptualizing a Design

The school board was well aware that behind every interesting design (as in our earlier examples of the Panama Canal, the *Whole Earth Catalog*, or the Cannes Film Festival) there is an interesting idea. Thus the first step in the process was for the design task force to come back to the school board with a proposed design concept—not an actual design—that would capture the imagination of educators and the community alike. The concept would be backed up by a cogent rationale, aimed at leading quickly to consensus. However, no further steps would be taken until a solid consensus was reached.

Because of the understandings gained during previous efforts in the district, it did not take long to reach consensus on some basic propositions. Ideas were drawn from personal experience, the reform literature, conferences, and consultants. The result was a design concept that emphasized coherence both in the development of students' understanding and in the thematic topics that organized study. The task force believed that the concept made sense as a whole, across the grades from kindergarten to high-school graduation and across the various subjects.

Reaching agreement on the importance of developmental coherence was relatively easy because participants in the design process believed that most of the significant ideas specified in their goals, had to be learned progressively over time. The Edmond Halley teachers were familiar with the growth-of-understanding strand maps found in the *Atlas of Science Literacy* and believed that it was possible, taking into account the findings of cognitive research along with the logical connections among ideas, to sequence concepts and their precursors across grade ranges. (Indeed, many of the district's science, mathematics, and technology teachers had long been looking to those strand maps for clues on improving coherence in their own teaching.) The design concept simply called for the developmental sequence and connections in the new curriculum to be explicit enough that responsibilities could be assigned and progress monitored.

Reaching agreement on thematic coherence was much more difficult because the idea was less clear than that of developmental sequencing, and because there were so many attractive possibilities. Eventually, after much districtwide discussion, the task force adopted "exploration" as the dominant theme for the Edmond Halley curriculum. The theme was meant to further the design of a curriculum that would engage students in investigating different aspects of the world and human existence from different perspectives: humanistic, scientific, economic, aesthetic, vocational. The curriculum would be characterized by the spirit of inquiry as manifested in different subject-matter areas and would build on children's natural curiosity about themselves and the world and its things and events. The curriculum would draw on the basic disciplines and fields.

The task force made clear in describing the design concept that the curriculum should be designed to do more than simply preserve students' curiosity and inquisitive behavior. It should also help them channel their curiosity so that they could explore their biological, physical and cultural surroundings more and more effectively as they matured and could continue to do so as adults. The arts, humanities, and sciences each have their own traditions, and students should experience exploration in each of those domains, not to become competent practitioners in them but to understand the nature of inquiry in each. The task force argued that the curriculum should engage students in exploring the kinds of explanations given by different disciplines, groups, and cultures, and at different times in history, for why things are the way they are. The exploration theme was also taken to mean that the curriculum should give students opportunities to probe the different occupations—not only the "professions"— and what it would take educationally to enter them.

The idea of exploration was intended to imply certain limitations and demands. For one thing, "exploration" implies a level of understanding well short of "mastery." The curriculum should, for instance, enable students to explore aspects of scientific inquiry (or artistic expression) without expecting them to master the art of conducting actual scientific investigations (or becoming skilled in art). As the task force pointed out, the demand implied by exploration in science is that students develop the basic mathematical and language skills as early as possible, for without them they will be severely handicapped when it comes to participating meaningfully in the exploration of any of the domains of human thought and action.

Selecting Curriculum Blocks

Moving from general propositions to concrete specifications was not a matter of simple logic for the task force. It was important for the process used to be interactive and unhurried, making it possible to examine a wide range of suggestions thoughtfully, make arguments and counterarguments, negotiate trade-offs, and build further consensus. Cross-subject, cross-grade curriculum committees and subcommittees were established at every school in the district to support the task force, and representatives of each set met regularly—in person and on-line—to present ideas, argue their merits, bargain, and assess progress.

The journey from design concept to design specifications was complete when the committees and subcommittees had managed to reach a strong districtwide consensus that was then approved by the task force and accepted by the school board. Hence an up-or-down vote between competing designs was never needed. Perhaps that happened in part because everyone understood that agreeing on design specifications was not the end of the line, that there were further stages to go through during which the specifications could be revisited and then, if they were not leading to an acceptable kind of design, revised. The task of establishing selection criteria for blocks was divided into two parts, one to focus on learning outcomes and one to look at the other properties a satisfactory curriculum would need. The results of these efforts are summarized below:

Goal specification. The task force concluded that it wanted the new curriculum to reach two broad curricular goals, one pertaining to all students and one serving the needs of some students. The all-student goal would be a common core that all students would be expected to learn; the other goal would be to offer to some students learning opportunities that go beyond the common core. The all-student, or core, part of the curriculum would include studies focusing on achieving literacy in the arts, humanities, science, technology, and mathematics. In the early grades, the core studies would emphasize developing the basic language and mathematical competencies needed to enable students to make continuous progress in them, while at the same time whetting their appetite for exploring the subjects further in the later grades and beyond. The responsibility for developmental and thematic coherence was to rest with the core curriculum, which (unlike the elective curriculum), had to be designed as a whole and linked to an explicit set of specific learning goals.

The task force sought authoritative content recommendations, having no intention of taking on the formidable job of deciding what would be most important for everyone to learn. This proved easy to do because the task-force members had been using the relevant documents in their own subject areas over the years. Recommendations were forwarded to the school board in every subject area. For science literacy they recommended that the core curriculum aim at all of the growth-of-understanding goals in *Benchmarks for Science Literacy*. These goals were adopted formally by the Edmond Halley school board and then forwarded to and approved by the state. There was now a clear understanding up and down the line that any curriculum design satisfactorily serving those benchmarks would meet national, state, and local content standards in science, mathematics, and technology (and similarly in other academic areas).

Constraint specification. Inevitably, the curriculum design had to accommodate some constraints. There were rules and regulations set by the school board and state law covering such matters as the budget for instruction-related expenditures, the length of the school year and day and its permissible subdivisions, and the assignment of teaching responsibilities. Because of its commitment to coherence, the task force set a severe constraint with regard to the division of time between the core curriculum and the curriculum electives. It believed that the existing curriculum was too much of a hodgepodge—students were free to choose unwisely and could end up being inadequately prepared for either work or college and lacking basic literacy in the arts, humanities, or sciences. Yet the task force also wanted enough flexibility in the curriculum to enable students to follow their interests and talents beyond the core, and it believed that the right kind of curriculum ought to enable all students to meet the core learning goals by the end of the 11th grade. So the task force indicated that, except for the senior year of high school, no more than a quarter of the time available could be devoted to electives.

Selection rules. With goals and constraints in mind, the task force developed specific requirements for a design. Chief among them were the following:

- Because the district had already adopted variable scheduling, curriculum blocks could be a semester or one to three years in length, and they could meet two, three, or five times a week, and each meeting could be one or two periods long. However, the final array of blocks would be required to form a pattern that students and teachers could follow easily.
- The exploration theme would be addressed by having a mix of discipline-related and integrated issue-related blocks, with some of each being in project format.
- To ensure that coherent understandings would result, at least one capstone block cutting across the arts, humanities, and sciences should be introduced at the end of the middle grades and again by the end of the 11th grade to review and organize the studies of the previous three or four years.
- To ensure that the design would be developmentally coherent with regard to science, mathematics, and technology, the core blocks selected for each grade range (K-2, 3-5, 6-8, and 9-12) would collectively target all benchmarks for that range.

Completing the Design

In principle, many different assemblies of curriculum blocks could meet the design criteria, but there was no reason to believe that all possibilities would be equally acceptable. Thus, as happens in nearly all design situations, the participants thought it important to consider a number of alternative designs.

Three districtwide design teams were appointed to propose separate designs. They were helped in this task by a curriculum-design consultant who gave the teams descriptions of curricula developed in other school districts that might meet many or all of the district criteria, and by examining a database of curriculum models. Each team was free to base its design entirely on an existing model or on a modification of an existing model, or to create one from scratch.

All three teams chose to adopt the same curriculum model—but each team modified it somewhat differently. The main part of their work then became the search for an appropriate arrangement of curriculum blocks. Each used a computer program to identify candidate blocks, to investigate and compare them, to assist in placing them, and to track how well the developing array satisfied targeted goals and necessary constraints. In making choices, each team had to trade some goals off in favor of others.

The three alternative designs were submitted to the task force responsible for overseeing the entire design process and making a final recommendation to the Edmond Halley school board. It had taken a long time to reach that point, and the work involved a very large number of people, but that is what is to be expected in important and politically difficult design undertakings. (For example, the design board for the United Nations headquarters in New York City met 45 times and considered 86 design concepts before reaching a final decision.) The new curriculum was coming into being in a way that would make its eventual adoption and successful implementation less problematic, even though it could be considered quite radical compared to the curricula of the 1990s and earlier. But there was still more work to be done.

The Edmond Halley task force studied the three competing designs. It debated the issues associated with them, including equity, cost, risks, possible side effects, manageability, adherence to state policy and legal requirements, and more. As it happened, however, the committee did not recommend any one of the designs as submitted. Rather, it made some trade-offs among them and settled on a compromise design that incorporated complementary features of each. The final design included required blocks and optional ones (selected from the three submissions) at each grade-range level, giving schools the authority to make some choices.

Lewis & Clark Regional School District

Meriwether Lewis and William Clark are adjoining rural counties. At one time, each had its own independent school system, but as academic demands increased and the local economies languished, the citizens decided that it made sense to pool their limited resources, even at the cost of local pride. The counties formed a single K-12 school district, aptly named the Lewis & Clark Regional School District. Not surprisingly, the merger raised difficult legal and political issues—such as how taxes would be levied fairly, where the one high school that would serve both counties would be located, what the makeup of the new board would be, which county superintendent would head up the joint system—as well as strictly educational ones.

In the turmoil of striving for the compromises and trade-offs needed to make the merger possible, a realization grew that this was an excellent time to rethink the curriculum, since, in more settled times, change of any significance had always seemed beyond reach. It dawned on the two communities that cost-effectiveness was not the only reason for the merger, maybe not even the most compelling reason. What finally became widely accepted, especially among parents in the two counties, was that the crisis before them was educational: their children were being left behind in the new world of global competition driven by advances in science and technology because they were not being educated for such a world.

With funding from the National Science Foundation's Rural Systemic Initiatives program and the loan of some corporate executives and technical personnel, a working design committee was commissioned to design a curriculum that would prepare the school district's graduates to realize their full potential and be the match of their peers nationwide. The design committee was encouraged to consult widely beyond the counties' boundaries, but to keep firmly in mind the geographical, cultural, and economic realities of life within the counties.

Settling on a Process

One of the first realities to be faced was that people were widely spread out and getting together for frequent meetings was very difficult. For that reason, the first decision on process was that the curriculum committee would consist of teachers, parents, and citizens in equal numbers from both counties but would hold its meetings in a relatively small area where the counties abut. The committee would meet as often as it needed to make progress, and it would report to the community at least once a month, with each report to be followed by public responses to questions and comments.

The participation of teachers was to be more organized. Since every school in both counties had computers, agreement was easily reached to link them with a server dedicated to the design challenge (upgrading some of the computers, as necessary). This necessitated hiring full-time computer specialists, but the school boards knew that that was an investment long overdue anyway. With the two-county computer network in place (and not to be used for instruction), schools could communicate easily with each other and with the curriculum committee and could access outside ideas and information by means of the Internet (or its 21st century equivalent). Moreover, the school board members could follow the discourse as it took place, reducing the need for the committee to make frequent and detailed reports to them.

The plan was for the committee to proceed one phase at a time. It would do some work, present its ideas on the dedicated network, consider the responses, revise the ideas as necessary, present the revised ideas, consider the responses, and so on until enough agreement had been reached to enable the committee to move on to the next step. In other words, most of the initiatives would come from the committee, but those teachers and administrators who wished to do so would be able to influence the outcomes on a daily basis. And neither the committee or the teachers could get too far without public input. The new single school board had to approve the key decisions.

Conceptualizing a Design

No one doubted that, given the opportunity, children in the two counties were perfectly able to learn the science and mathematics that children anywhere could learn. What the school boards and curriculum committees did doubt was their ability to attract the quality of teachers and to offer the array of courses needed to provide such an opportunity. Geographical remoteness and financial circumstances made it hard to recruit outstanding teachers, and the small population of students in the upper grades, along with the ever-present financial limits, made on-site course variety unrealistic.

Facing these realities, but unwilling to give in to them, the committee proposed a curriculum concept that (with modifications along the line) was rather quickly subscribed to by teachers, students, parents, citizens, and the school board even before all of the details of the merger had been hammered out. The concept had three facets:

- In each major subject area, students would individually be held to the attainment of national or state standards, whichever was, in each area, higher.
- The skill to be developed above all others would be the ability to engage in independent learning.
- State-of-the-art information and communications technology would support the curriculum.

Briefly, the argument for this high-tech curriculum design ran something like this: Without challenging external learning goals, it would be too easy to settle for whatever locally defined goals that could easily be met. Moreover, since valid computerized tests keyed to state and national standards were known to exist, it would be easy to monitor the progress of each student. If, in the process of acquiring a comprehensive base of knowledge, students learned to be competent learners, they would then be in a strong position for the rest of their lives to go as far beyond the standards as the times and their interests urged. By depending heavily on the use of modern computer and communications technologies to build that base of knowledge, students would enhance their developing learning skills and gain the technical facility needed to make good educational use of the information highway.

Calculations indicated that purchasing, installing, maintaining, and updating the requisite technologies (including laptop computers), paying line charges, and training teachers and students in their use would be much less costly than what it would cost to attract teachers from the best universities and to run small classes of academic subjects. There were four basic assumptions behind the calculations: (1) help in underwriting the capital costs would be forthcoming from government and industry; (2) much of the training on the use of the computer system, and some of the maintenance, would be carried out by middle-school and high-school students; (3) many of the current teachers could be converted from "tellers" into learning guides, and preference would be given in future hiring to those teachers best prepared to help students learn to learn; and (4) there were outstanding curriculum blocks available—outstanding in their focus on learning goals specified in national and state standards—that were technology intensive and explicitly designed to foster independent learning skills.

Selecting Curriculum Blocks

Curriculum design proceeded straightforwardly from the design concept: assemble blocks, preferably high-technology, to create a curriculum that would enable students to reach challenging learning goals while becoming independent and technologically proficient learners.

Goal specification. As noted at the beginning of this story, the Lewis & Clark counties embarked on this adventure only after a decade of working in one way and another to improve the curriculum at hand. Part of that decade-long effort engaged teachers, administrators, and some interested parents and other citizens in becoming familiar with authoritative statements of learning goals in various fields, including the science, mathematics, and technology education standards. Because *Benchmarks for Science Literacy* (in its successive editions) substantially overlaps all of those and the state's framework in those subjects as well, and because *Resources for Science Literacy* (in its successive editions) provides a convenient way to move back and forth among them, the learning goals specified in *Benchmarks* were adopted without modification.

Constraint specification. To purchase and maintain the advanced computer system that the new consolidated district would rely on, the district would have to forego hiring top-flight content specialists for every classroom in the upper grades (even if they could be attracted in sufficient numbers, which was itself doubtful). As a consequence, the new curriculum could not include courses that required such teachers. Similarly, blocks that required special facilities, such as laboratories for each of the sciences, would not be eligible. Instead, the new high school would have one large project room for science that could be used by individuals and small groups for hands-on activities associated with their studies. It would also house a reference library of textbooks, from easy to advanced, in science and mathematics subjects. However, no textbook adoptions would be made, nor would any course be offered that required all students to have the same textbook.

Finally, block selection would minimize establishing a rigid sequence of courses, especially after 8th grade, since students would be permitted to pass courses by examination and in any order. Indeed, no courses as such would be required, though students could form study groups to explore together (with a faculty guide) any of the required subject areas. For that purpose, the faculty would identify interesting and goals-relevant blocks that it would be willing to lead, given enough student demand. Also, students themselves—individually and as small groups—would be free to browse the database of curriculum blocks to find ones for consideration by the faculty that would move them toward their personal learning goals, as well as those set by the combined school district.

Selection rules. Translated into practical terms, the criteria for assembling the Lewis & Clark K-12 curriculum became:

- Every block for grades K-2 and grades 3-5 had to be rich in benchmarks and be teachable in an ordinary elementary school with a generalist teacher. The blocks for grades 6-8 could either each have some provision for independent study, or have among them some that did not include independent study and some that were entirely independent study. The high-school curriculum would consist largely of a pool of independent-study blocks.
- There had to be one common block in high school (taken in parallel with whatever science, mathematics, and technology students were studying individually or in small groups) that dealt with the nature of the scientific enterprise. It could be any length, from a semester to two years, provided it met for a total of at least 180 class hours.
- The curriculum had to include some "teaching blocks" that would prepare middle-school and high-school students to share in some of the teaching of reading and computation skills in the lower grades. Priority would be given to blocks focusing on the use of calculators, computers, and networks.

- Increasingly from kindergarten through graduation (which could come anytime after a student had passed the requisite examinations), curriculum blocks had to be technology dependent. That could include the entire array of electronic and print media, but not conventional textbooks alone. Blocks could be selected that referred to standard textbooks, but not blocks built around a single, required textbook. Preference would be given to blocks referring to trade books and articles available in an ordinary school library or that could be obtained by loan from a distant library or on-line.
- Since specified learning outcomes were the basis for progress, all blocks had to contain appropriate evaluation instruments and procedures. That was considered crucial, since students would be expected to monitor their own learning and teachers would be expected to monitor how well students were able to do so.

Completing the Design

Because of the limited transfer of students among elementary schools in this region, there was no special interest in having all of the elementary schools end up with the same array of blocks in their curricula. The board approved, nevertheless, having all students take the same tests at the checkpoints, and basing the tests on the district's specified learning goals rather than on the individual school curricula. It was also agreed that each student's passage from 5th grade to 6th grade and from 8th grade to 9th grade would depend upon the student demonstrating the technology skills that would be required in the middle and high schools, respectively.

Each county had just one middle school, and the committee's task was to come up with a curriculum that would make up for any lapses in the earlier grades and prepare students for the one high school serving both counties. They looked for blocks that would help students make the transition from dependent to independent learners. Most of the blocks were designed around small-group instruction, with some individual study, and only a few were in the form of whole-class instruction.

For several reasons, the high-school curriculum was more difficult to formulate. For one thing, the two schools had formerly been rivals, at least in sports. What was more important was that neither had had much experience with goals-directed instruction, independent study, or advanced technologies. And the new high school with adequate space and wiring did not yet exist—though there was the chance to opt for a school to match the curriculum, rather than the reverse.

The combined faculties finally decided to move cautiously by creating the curriculum over a five-year period. During the first year, they worked out the 9th-grade curriculum and installed it in the second year (all other students in year two took the traditional curriculum, more or less). On the basis of that experience, the 10th-grade curriculum was designed in year two and instituted in year three, and so forth. It turned out that as faculty and students gained experience and confidence, and with substantial faculty turnover due to retirements, the transition was completed in four years. An interesting consequence was that the design process—students and faculty consulting in one year on the composition, within limits, of the curriculum for the following year—survived and became a permanent feature of the Lewis & Clark high school curriculum.

AFTERWORD

The three stories above constitute a brief look at the process of designing a curriculum by assembling components according to agreed-upon selection rules, not a description of what the curricula of the Palladio Unified, Edmond Halley, and Lewis & Clark Regional school districts would have turned out to be—or what modifications of them would have resulted from monitoring how they worked. Perhaps a central point to be made is that different school districts can approach the design challenge in quite different ways administratively even while following good design practice and drawing on the same pool of possible curriculum elements. The stories also suggest—even though they skip over the final designs of each of the three districts—that it is possible to create very different curricula to serve a common set of learning goals.

Before such stories can become a reality, the requisite curriculum blocks and the computerized curriculum-design system will have to be created (all design challenges in their own right). Perhaps the promise of a new age of curriculum design and development suggested here is enough to motivate us individually and as a nation to get busy on those tasks and design our own nonfictional curricula.

In our three stories, the school districts were remarkably fortunate in being granted ten years to get ready for curriculum reform. Although in reality, none of their design endeavors would have run as smoothly as portrayed here, virtually free of personality clashes and disagreements, even their idealized success could hardly have been contemplated without that decade of preparation. What would they have done in that decade of grace? Part III describes some of the essential strategies of how to get started.

PART III
IMPROVING TODAY'S CURRICULUM

Eventually, say in a decade or so, all the pieces may be in place for schools finally to be able to undertake radical curriculum design in a thoughtful, informed, and systematic way, as described in Part II. But there is no need to ignore the present to serve the future. The work of Project 2061 is based on the premise that it is possible to make significant improvement in the current curriculum while pursuing a long-term reform strategy—as long as what is done right now is not a mere palliative. The greater danger is to ignore the future while serving the present.

The following three chapters suggest steps that a school district can take now to improve its K-12 curriculum. Because the recommendations for getting started involve teachers gaining experience in a variety of new strategies and techniques, and developing skills for making curriculum decisions in collaboration with other teachers, all three chapters can be characterized as "building professional capability" even though only the first chapter has that title.

CHAPTER 6: BUILDING PROFESSIONAL CAPABILITY discusses kinds of professional enhancement that can put educators in a position to make meaningful improvements in the existing curriculum, using a set of tools already developed by Project 2061.

CHAPTER 7: UNBURDENING THE CURRICULUM suggests how to make room for teaching the most important concepts and skills by eliminating less important topics, pruning topics of unnecessary detail, trimming technical vocabulary, and ending wasteful repetition.

CHAPTER 8: INCREASING CURRICULUM COHERENCE outlines ways to improve a curriculum by taking into account the development of understanding over time and by connecting concepts within and between subject-matter domains.

The three chapters do not need to be read in the order given here; ideally, the kinds of reform activities described in them should all be under way at the same time. Moreover, after completing Part III, some readers may want to return to Part II to consider again how design in the future would depend on the getting-started activities described here.

For understandable reasons, educators want to get started on actually improving the instruction of the students they are now teaching. Their emphasis is on action now—doing something to the curriculum or some part of the curriculum, in contrast to deliberating about it. But action, without carefully drawn aims, without enough thought and information, without suitable preparation for those who must make change happen, and without a means to estimate its effects is unlikely to pay off in the long run—however satisfying it may be in the short run.

It is important, therefore, that careful thought be given to the process by which decisions are made and carried out before initiating lots of changes. In other words, it matters as much how reform is approached as what steps are taken. Suggestions for those approaches appear frequently in these three chapters and can be summarized briefly:

- Engage as many stakeholders as possible.
- Form cross-grade, cross-subject teacher committees to participate in curriculum design.
- Work toward a shared commitment to learning goals, but do not suspend action waiting for a complete consensus.
- Relate the local reform vision to national goals.
- Keep accurate records.
- Proceed in the spirit of science.

Below we offer a more extended discussion of each of these points. However, some readers may wish to go directly to Chapter 6 instead and then return to this discussion when it becomes relevant to their thinking about process.

Engage as many stakeholders as possible. This is a standard recommendation, but one that is easier to proclaim than to carry out. Though it can be a burden and slow things down, getting stakeholders involved is a price well worth paying. The reason most often cited for broad participation is that curriculum changes, however well conceived, are likely

In his book *Engineers of Dreams* (1996), Henry Petroski shows how important external reviewers can be when it comes to bridge design; external review cannot be less important in the more complicated business of curriculum design.

to remain isolated in a few schools, rejected out of hand by those teachers, administrators, and concerned citizens who are left out of the design process. All should at least be invited to participate whether they eventually elect to do so or not. But mere invitation is not enough. Participation will be much less convenient for some stakeholders than for others, and it is important to persistently encourage and facilitate participation by them as well.

There are, of course, reasons other than political strategy and democratic principles for seeking widespread engagement. Quality will be materially improved by having teachers, administrators, parents, students, and interested citizens in a school district participate in some way in the design process. Engagement requires everyone who is involved to reconsider what is fundamental in education and what is not, and it can help them to move away from slogans and easy generalities to more thoughtful notions of what students should learn. When so engaged, educators and scientists (and other scholars in the cases of other subject domains) learn to work together for a common cause; K-12 teachers learn to collaborate across grade levels and subject-matter boundaries; and the more individuals there are who understand and subscribe to the goals, the more chances there are that a design eventually can be agreed upon and followed. None of this is meant to guarantee that there will be no conflicts, but conflicts aired early tend to cause less trouble than ones that are ignored or deferred.

It is not likely to be practical to include members of the public in the more detailed aspects of design work. Rather, their engagement should be in advisory committees that generate enthusiasm and support, approve of plans for recruiting and organizing teacher teams, review progress, and eventually promote the results to their own constituencies.

Form cross-grade, cross-subject teacher committees to participate in curriculum design. To the degree possible, the committees should avoid grade and subject stratification, which almost always promotes piecemeal change. Making changes in the elementary grades without regard for the consequences in the middle grades is likely to lead to problems when students advance from one level to the next. Changing the content of science courses without taking advantage of changes made in mathematics courses could result in a lost opportunity to do a better job in both.

The fact that schools are geographically separated from one another by grade level makes K-12 collaboration difficult, but not impossible. In many urban school districts, for instance, informal discussions can be set up among a high school and its feeder schools. And any number of schools in a district can be connected by a computer network that enables them to interact quickly and continually.

Work toward a shared commitment to learning goals, but do not suspend action waiting for a complete consensus. Except at the most abstract levels of generality, consensus on content goals is not easy to reach. Curriculum committees should use what agreement they have as a continuing reference point, asking themselves how any proposed action will contribute to those ends and requiring members to make arguments only in terms of likely effects rather than philosophical preferences. Dissension can arise, of course, but this process stands a chance of gradually enhancing everyone's understanding of goals and increasing the relevance of the conversation. This is not to suggest that school curriculum groups actually create specific learning goals from the ground up. They lack the resources to do a credible job of it, and it is not the best use of their time.

Relate the local reform vision to national goals. Fortunately, there is no need to start from scratch. In most fields, scholars and practicing educators have been convened at state and national levels and provided with resources—including the necessary time—to recommend what students should end up knowing and being able to do. Moreover, those efforts usually had to pass many levels of expert and public criticism. By starting with these, but still taking the time to argue their way through them, committees can get a head start on achieving consensus on learning goals. But beginning with carefully worked-out national goals is still not easy. Such goals in science, mathematics, and technology go considerably beyond mere lists of topics, to specify more or less precisely what is to be learned. Using them well requires careful study of their specific meaning.

Suppose, for example, a school district subscribes to the general notion that all students should become science literate by the time they graduate from high school. There is no need for it to try to decide on its own what constitutes science literacy, since *Science for All Americans* sets out a vision of science literacy that has been widely supported by scientists and educators and has helped to shape national and state standards. As participating teachers become familiar with that vision, they will want more grade-by-grade detail, for which they can turn to *Benchmarks for Science*

Literacy, the *National Science Education Standards,* the mathematics standards of the National Council of Teachers of Mathematics, and the technology standards undertaken by the International Technology Education Association—or to state frameworks that have been conscientiously based on these documents. Reference to national and state learning goals has been made easier by the item-by-item comparisons of them made available on *Resources for Science Literacy: Professional Development.* In every case, of course, some adaptation may be needed to meet local conditions.

Keep accurate records. When analyzing the existing curriculum or working on the new design, curriculum committees should be sure that the data collected, conclusions reached, and actions to be taken are recorded where they can be recovered later. The design process calls for testing the effectiveness of key ideas and components along the way. If something is important enough to be put to a test, it is worth the time to analyze the results critically, to inform the design effort. Tests and their results should be recorded in the school district's growing curriculum-reform database.

Proceed in the spirit of science. This means looking upon actions taken to change this or that aspect of the curriculum as experiments and as opportunities to understand the curriculum more deeply. There is no need for every action taken in the name of curriculum design to succeed. If actions and results are monitored carefully, as much may be learned from what does not meet expectations as from what does. New formats or styles can initially be puzzling to both teachers and students. If there is good reason to believe that some innovation will work—for example, if it has been successful in other schools—then some time of adjustment may have to be allowed to see its benefits.

Of course, students should not needlessly be put at risk of learning less or of learning less important things than they would learn otherwise, or of learning whatever they learn less well. But the current curriculum has its own risks; indeed, research on learning shows that students typically learn much less than their teachers believe they do. So the question is not of risk versus no risk but of trade-offs among risks.

The purpose of the approaches suggested here is not to add complexity to reform undertakings. It is, rather, to increase the likelihood that action will be well thought out and well supported. The tasks at hand—achieving significant professional development, unburdening the curriculum to make room for the thorough study of the most important concepts and skills, and increasing curriculum coherence—are by their nature difficult. They simply cannot be successfully carried out quickly or by a few people. But in the long run—achieving sustainable curriculum reform is inescapably a long-run matter—more progress will be made by pursuing a systematic approach to curriculum reform.

Jack Boucher, *Shaker Guest House*, 1963

CHAPTER 6
BUILDING PROFESSIONAL CAPABILITY

The professional preparation of new teachers concentrates on getting them ready to teach certain subjects and grades in the existing curriculum, not on how to go about changing the curriculum they have inherited (let alone learning of the need to do so). Consequently, professional development of employed teachers tends to focus on improving their content background and instructional techniques. Similarly, the preparation of school administrators focuses largely on matters of school management, with little attention paid to the details of curriculum design.

If school districts are to achieve curriculum reform, therefore, it is essential that they build a professional capability for undertaking curriculum change. This chapter suggests how school districts can make headway in developing such a capability while at the same time beginning to make substantial improvements in some aspects of the curriculum itself. It calls upon educators in school districts to raise their collective level of science literacy, to become knowledgeable about the science, mathematics, and technology learning goals appropriate for all students, and to familiarize themselves with what is reliably known about student learning related to those goals. The chapter also calls for educators to increase their ability to make sound judgments about the quality of curriculum materials and to employ a variety of different curriculum formats that have particular instructional advantages.

Building professional capability for curriculum improvement in a school district calls for teachers to acquire certain knowledge and skills regarding science itself, how it is represented in literacy goals, how students learn challenging ideas, and how materials for instruction and assessment serve students' learning. It does not follow, however, that every teacher and administrator in a school district must attain the same level of expertise in every one of these matters. If one thinks of the faculty as a large team of professionals engaged in a shared endeavor, then it is the collective capability that counts—as long as there is sufficient collaboration among the members of the team. Teamwork is not common enough in the area of curriculum reform, but it is an aspect of professionalism that can be learned, if there is a will to do so. Therefore, the suggestions that follow are framed as team undertakings because the development of team skills is itself part of building professional capability.

INCREASING FACULTY SCIENCE LITERACY

Science for All Americans argues that all high-school graduates should be science literate, and it describes the science, mathematics, and technology knowledge and skills that constitute such literacy. Although, in principle, all teachers should have reached that same level of science literacy, the present-day blunt truth is that too few of them—and too few college graduates in general—have done so. Perhaps someday all teachers will be science literate when they enter the profession, but in the meantime, steps need to be taken to enable teachers in a school district to make substantial progress toward achieving science literacy.

All of the ways suggested here require individual effort and administrative support. Learning takes time, resources, and encouragement. Without recognition by the community, school board, and administrators that teachers must upgrade their subject-matter knowledge continually, and without policy and budgetary support for the needed time, resources, and encouragement, in-service professional development will be of little consequence and contribute little to building districtwide professional capability. But with such recognition and support, teachers can improve their understanding of science, mathematics, and technology by engaging in individual or group study of selected readings or growth-of-understanding maps, setting up a long-term program of workshops, or by taking appropriate courses.

Readings

Because it defines adult science literacy, *Science for All Americans* can provide a focus for faculty study. It does not serve well as a textbook, its purpose being to summarize rather than to teach, but excellent books and articles are available that cover in detail most of the topics in *Science for All Americans*. Project 2061 has undertaken the task

"A three-year study of education reform found that most staff development activities 'were too short and lacked the follow-up necessary to develop the deep content and pedagogical knowledge necessary to meet new instructional goals...[and] did not appear to be building an infracturcture to promote and sustain teacher learning and instructional improvement over the long term.'"
—"The Bumpy Road to Education Reform" in *CPRE Policy Briefs* (*June 1996*)

"Curriculum-reform efforts are hard to sell and even more difficult to sustain if they can only succeed if teachers have special capacities, such as: extraordinary subject-matter expertise; the time, will, and skill required to develop their own curriculum materials; the ability to teach widely divergent students effectively; and the ability to maintain control over these students while allowing them freedom to learn on their own."
—D. F. Labaree, "The Chronic Failure of Curriculum Reform" (1999)

of sorting through the tens of thousands of books marketed to the general public that deal with science, mathematics, and technology to find those that are the best for this aspect of professional development. *Resources for Science Literacy: Professional Development* is a CD-ROM containing five different kinds of resources linked to *Science for All Americans* (see next page). Among them is a compendium of what Project 2061 believes are some of the best books available for individual and group study. Each book is described, published reviews of it are reproduced, and its ties to *Science for All Americans* are specified.

Descriptions of recommended trade books can be found on *Designs on Disk* and on Project 2061's Web site at www.project 2061.org.

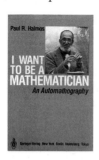
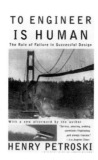

Although teachers can use the CD-ROM to design and pursue individual programs of study, group study should be encouraged. One approach is to form reading groups in each school. After agreeing on a topic, each group discusses the reading possibilities and then selects a book for everyone to read and discuss in subsequent meetings.

Study groups will differ in the number of participants, whether teachers from other schools and parents are invited to attend, how many books are taken on each semester, how sessions are conducted, and so on. Within reasonable limits, such variations are not likely to affect the outcome greatly. What is important is that the readings be selected by the group itself and that participating faculty can earn appropriate professional development credit. It is desirable, of course, that time for the group to meet be included in the formal school schedule. But the fact that scheduling practices in most schools often make it difficult for teachers to meet together during the school day need not be an impenetrable barrier. Each group should be able to find one evening, late afternoon, or early morning once a month on which to meet to discuss the reading. Alternatively, electronic conferencing can make it possible for members of a groups to participate in the conversation at their convenience. (Indeed, the group can use the World Wide Web to look for reviews and expert commentary on the book under discussion.)

Designs on Disk contains a database with convenient forms for recording reactions (positive and negative) to each book a study group reads. This process also enables the group to keep track of topics that have been studied by at least some members of the faculty and to identify teachers who are well informed on particular aspects of science, mathematics, and technology and can be called on as consultants by other teachers.

A revised edition of *Resources for Science Literacy: Professional Development* will also include a database of newspaper, journal, and magazine articles that shed light on topics that are central to science literacy.

This graphic menu from *Resources for Science Literacy: Professional Development* displays the contents of the CD-ROM.

CONTENTS OF *RESOURCES FOR SCIENCE LITERACY: PROFESSIONAL DEVELOPMENT*

SCIENCE FOR ALL AMERICANS
The full text of Project 2061's landmark report is available for the first time in an electronic format. Links to other components allow users to identify resources on the CD-ROM that are relevant to specific chapters and sections of *Science for All Americans*.

PROJECT 2061 WORKSHOP GUIDE
The Workshop Guide contains a variety of presentations, scripts, activities, and supplementary materials that can be used to design and conduct Project 2061 workshops.

COMPARISON OF *BENCHMARKS FOR SCIENCE LITERACY* TO NATIONAL STANDARDS
Detailed analyses compare *Benchmarks* to national content standards developed by the National Research Council, the National Council of Teachers of Mathematics, and the National Council for the Social Studies.

COGNITIVE RESEARCH
An introduction to current cognitive research literature, along with *Benchmarks* CHAPTER 15: THE RESARCH BASE and its accompanying bibliography of more than 300 references to the educational research literature, sheds light on how students learn particular concepts from *Science for All Americans* and *Benchmarks*.

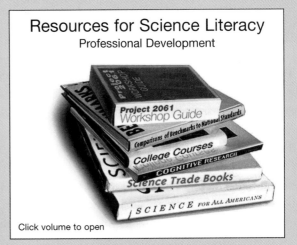

SCIENCE TRADE BOOKS
Full bibliographies, reviews, and other descriptive data are provided for more than 120 books for general readers dealing with many areas of science, technology, and mathematics. Each book is linked to related *Science for All Americans* chapters and sections.

COLLEGE COURSES
Descriptions of 15 undergraduate courses suggest how to teach college students particular concepts from *Science for All Americans*. The syllabi are linked to relevant chapters and sections of *Science for All Americans*.

Project 2061 Strand Maps

Study of these maps, which attempt to depict how students' understanding might develop over the school years, is a useful adjunct to a program of readings. The strand map example on the next page shows how the development of a concept can be traced from its simple beginnings as ideas join and grow in sophistication. Study groups may find it useful to work their way up a map, discussing its individual benchmarks and seeking information from the books on the reading list. Although the main purpose in studying such maps is usually to acquire a better understanding of the development of student learning or to plan curriculum sequences, many teachers have found that the process serves as an excellent organizing device for helping them improve their understanding of the topics—and to decide what ideas to teach and what to emphasize about them.

Courses

A time-honored way for teachers to acquire content knowledge and develop professional skills is to take college courses in science, mathematics, and technology. Many colleges and universities make such courses available on campus during the summer and academic year, but the institutions may not be within geographical or financial reach of teachers. And even if within reach, they may not offer the content courses that the teachers need to make progress toward science literacy. Once a teacher has reached a comfortable level of literacy in a particular area, regular college courses may be useful. The priority, however, should be on achieving literacy efficiently.

Colleges and universities are becoming more willing to tailor the content, instructional style, and scheduling of courses to fit the specific needs of groups of teachers. As a result, taking courses may become an important component of a school district's overall plan for building its professional capability for curricular reform. The concomitant increase in video and/or computer-based courses available at a distance can enhance that effort. Some universities are developing courses that will serve the science literacy needs of teachers, both preservice and in-service, by focusing on the image of literacy portrayed in *Science for All Americans* and accommodating the research on the prevalence of misconceptions in many areas. Such courses can improve not only the teachers' grasp of science, mathematics, and technology, but how those subjects can be taught effectively.

The use of some combination of reading groups focusing on content, study sessions built around strand maps, and courses and minicourses (summer and academic-year,

The syllabi of some tailored courses can be found on the *Resources for Science Literacy: Professional Development* CD-ROM, and more examples will be added in subsequent versions. Perhaps these syllabi will motivate many other colleges and universities to contribute to the database—especially if there is a clear demand for such courses from groups of teachers in a school district.

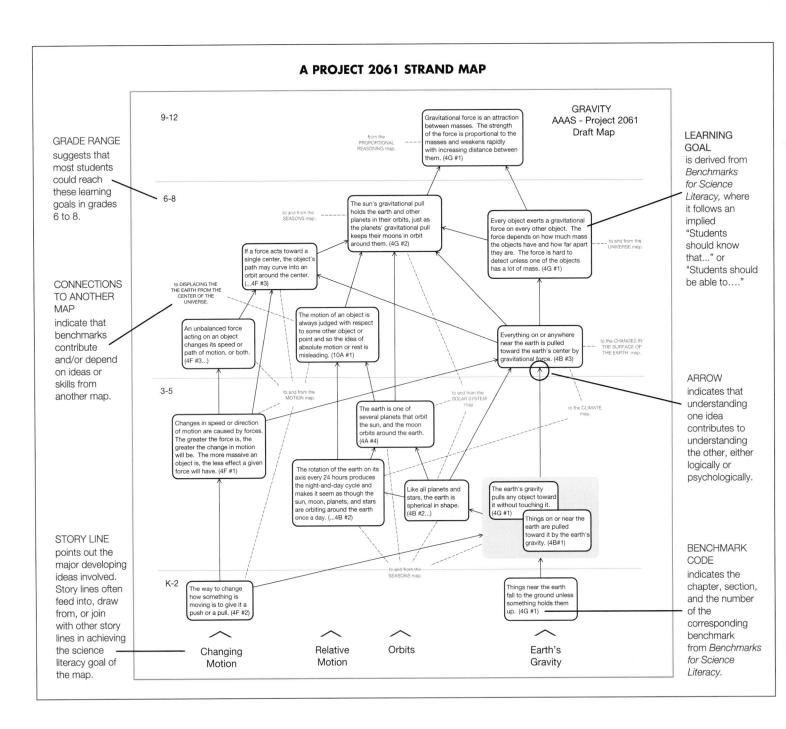

A PROJECT 2061 STRAND MAP

direct and by way of the Internet) can gradually lead to an increase in the proportion of a school district's faculty members who are science literate (as spelled out in *Science for All Americans*). With careful planning and some luck, summer research opportunities offered by local business and industry can contribute as well. In many locations, national laboratories also provide programs for teachers. Because of the years of service that lie ahead of younger teachers, and the need for them to have the habit of continuing education become ingrained, it is particularly important that they be included in this professional-development process. Simultaneously, school districts should raise their hiring standards for new teachers by stating explicitly that evidence of science literacy will be taken into account when hiring and by notifying the relevant teacher-education institutions of their expectation that candidates be science literate.

UNDERSTANDING STUDENT LEARNING GOALS

Having a faculty that is well grounded in science, mathematics, and technology is not enough to ensure that all (or even most) students will learn enough in school to become science literate. Research studies have shown that teachers' subject knowledge is only part of the story of successful learning. Equally important is their understanding of precisely what it is that they expect students to learn, the developmental pace at which students are able to learn those things, and the difficulties that students typically encounter.

Fortunately, a faculty does not have to determine for itself what appropriate student learning goals are for science literacy. The efforts of the nation's scientific and science teaching organizations over a period of years have resulted in publication of *Benchmarks for Science Literacy, Curriculum and Evaluation Standards for School Mathematics*, and *National Science Education Standards* (*NSES*). These reference works are in general accord on the importance of reducing the mass of an overstuffed curriculum and specifically on what science, mathematics, and technology knowledge and skills are most important for students to learn. Even if they are empowered to create their own science-literacy learning goals, local groups will find it valuable to study these reference works in detail (as distinct from just comparing topic headings).

None of the national groups has merely made a selection of assorted topics. They have attempted to identify interconnected sets of ideas and skills that will, in the *Science for All Americans* phrase, "maximize students' ability to make sense of the world and to learn more about it." Reformers should take care not to disregard the coherent set of specific learning goals in the national documents or to simply pick and choose

Also available: *Technology for All Americans,* a report of the International Technology Education Association, spells out learning goals in technology education.

For a more detailed discussion of the term "topics," see CHAPTER 7: UNBURDENING THE CURRICULUM.

casually among them. By so doing, they may lose not only important interconnections within or across topics, but also the potential for K-12 continuity that helps students to gradually build their understanding of difficult concepts. Reformers should also beware of simply *adding* national-goal recommendations to the requirements of an already unwieldy curriculum. The national goals for science literacy are designed to help educators focus on fewer, but more important, ideas so that all students have a chance to learn them well.

Understanding the real intent of a set of specific learning goals is not as straightforward as it may seem at first glance. The difficulty comes from taking the benchmarks and standards to be lists of "topic headings" (as is often the case with familiar curriculum guidelines), rather than as painstaking selections and specifications of the essential aspects of ideas to be learned and understood in relation to one another. For example, seeing the section heading "Cells" in *Benchmarks* could be taken as an endorsement to teach anything whatever about cells—including over a hundred technical terms typically found under the topic of cells in the high-school biology course—rather than the carefully chosen ideas that *Benchmarks* describes.

"Topics" have another muddling effect besides excessive inclusiveness. They often identify what is to be studied, without specifying just what is to be learned. As noted in CHAPTER 4: CURRICULUM BLOCKS, "acid rain" is a likely topic for a middle-school science unit. Neither *Benchmarks* nor *NSES* includes acid rain as a high-priority component of science literacy. Nonetheless, studying the topic of acid rain could help students toward any number of benchmarks that are high-priority components having to do with differences in climate, the mechanics of the water cycle, the appropriateness of measurements, fitting data with mathematical models, proportionality of concentration, the difficulty of anticipating side effects of technology, uneven benefits and costs of trade-offs, and so on. From the perspective of specific learning goals, acid rain is a *context* in which many such benchmarks can be pursued. The crucial distinction between what is to be learned (specific learning goals) and what is to be taught (topical context) is often lost in education discourse, with the former being taken erroneously to be synonymous with the latter. It is essential to keep the distinction straight.

How, then, can a school district foster the needed understanding of student learning goals among its faculty? One practical approach, framed here with reference to the science literacy learning goals set out in *Benchmarks,* is to use Project 2061 tools: to study the growth-of-understanding maps; analyze instructional topics against specific learning goals; and participate in the kinds of the workshops described on

Resources for Science Literacy: Professional Development. The next three sections describe these three activities.

Studying Strand Maps

Teachers report that the Project 2061 maps about students' growth of understanding are especially valuable when used by small study groups of elementary-, middle-, and high-school teachers all working together. Larger groups can form small subgroups and have them compare their interpretations from time to time. A common way to proceed is to follow three basic steps:

First, the group should start with maps on familiar topics that most of the group members feel comfortable with, and then move on to those perceived to be more complicated or less familiar as the group's confidence builds. "The Water Cycle," "Culture and Heredity," and "The Conservation of Matter" are generally well received by teachers undertaking the study of growth-of-understanding maps for the first time.

Second, the group should generally work from grade K-2 benchmarks toward ones for grade 9-12, although there is likely to be a lot of back and forth. Considering one benchmark at a time, the group members should first discuss what they think it means and doesn't mean. They should read the appropriate section of *Science for All Americans*, then the essay in the section of *Benchmarks* in which the benchmark under study is found, and then the relevant research findings, if any, cited in Chapter 15 of *Benchmarks*. Then the group should once again discuss the benchmark.

Third, the group should run through the map again, this time discussing the connections among some of its benchmarks. The group members should consider what each arrow means—whether it indicates a necessary prerequisite or only a helpful contribution. Seeing the relationships between two or three benchmarks will help to clarify the meaning of each of them and to make explicit what is meant by growth of understanding.

Analyzing Instructional Topics

Faculty study groups should analyze instructional topics in the light of specified learning goals. Briefly, there are three steps to be taken:

First, the group should identify a contextual topic (for example, "Cloning") that it believes offers a rich opportunity for learning, and then try to agree on which benchmarks, if any, could be targeted in studying that topic at each of the indicated grades.

Second, referring to adopted textbooks, course outlines, and curriculum frameworks, the group should check the list of specific learning goals (benchmarks) against

Benchmarks on Disk, Resources for Science Literacy: Professional Development, and the Project 2061 Web site include several sample maps and commentary on them. Many more are scheduled to appear in *Atlas of Science Literacy.* Further suggestions on using maps appear in CHAPTER 8: INCREASING CURRICULUM COHERENCE.

CLARIFYING A PARTICULAR BENCHMARK

In Project 2061 workshops, the full meaning of a benchmark is revealed through careful consideration and study of the following:

- **Adult Literacy Goal.** For each *Benchmarks for Science Literacy* section, there is a corresponding *Science for All Americans* section describing adult science literacy goals for that topic; it can help participants understand where benchmarks in that section are aiming.
- **K-12 Context.** A review of benchmarks for other grade levels from the same *Benchmarks* section helps participants understand the level of sophistication intended by the benchmark.
- **Instructional Strategy.** The introductory essays in the *Benchmarks* section for the benchmark being studied help participants understand difficulties students may have with the concept or skill and offer some suggestions for helping students achieve the benchmark.
- **Research Base.** Summaries of research on the topic from *Benchmarks* CHAPTER 15: THE RESEARCH BASE suggest likely limitations in student understanding of the benchmark and, therefore, imply an appropriate grade level for the benchmark. They also point participants to the original research articles.
- **Strand Maps.** A relevant conceptual strand map depicts K-12 growth of understanding for a particular *Science for All Americans* idea. The maps on which a benchmark appears help participants see how other benchmarks relate to and contribute meaning to the benchmark being studied.

There is no special order to these activities, although elementary teachers often prefer to see where an idea leads next in strand maps, whereas high school teachers often look to the eventual context of understanding in *Science for All Americans*.

Consider the grade 6-8 benchmark on the flow of matter and energy: "Food provides molecules that serve as fuel and building materials for all organisms...." Educators who read the precursor grade 3-5 benchmark, "Almost all kinds of animals' food can be traced back to plants," are more likely than those who don't to realize that the grade 6-8 benchmark goes beyond "what eats what." By reading the essay, educators learn that following matter through ecosystems needs to be linked to study of atoms, which itself is impractical before late middle school. Research alerts educators to the difficulty students have accepting the idea that plants make their only food from water and air. And the strand map Flow of Matter and Energy shows how understanding the conservation of matter in living systems depends on understanding the structure of matter.

what is actually taught. The group's task here is to spot where mismatches seem to occur—such as aspects of the topic being taught that are unrelated to any of the selected benchmarks, or instruction having no bearing on some of the benchmarks, or instruction relating to some of the benchmarks being positioned earlier or later than those benchmarks. The members of the group should review the selected benchmarks to make sure they understand them, using the techniques described in the box on the facing page and the one below. They can then modify the topic/goal list to obtain a statement of learning goals for the topic that teachers at all grade levels can support.

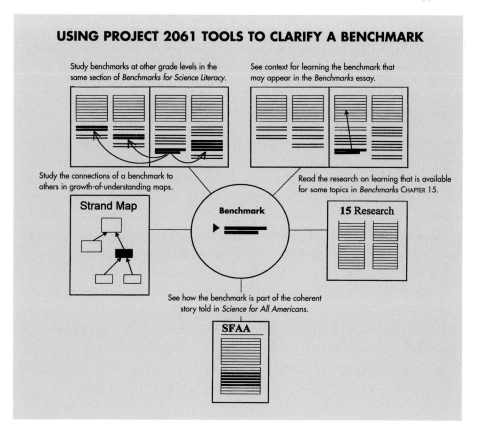

USING PROJECT 2061 TOOLS TO CLARIFY A BENCHMARK

Study benchmarks at other grade levels in the same section of *Benchmarks for Science Literacy*.

See context for learning the benchmark that may appear in the *Benchmarks* essay.

Study the connections of a benchmark to others in growth-of-understanding maps.

Read the research on learning that is available for some topics in *Benchmarks* CHAPTER 15.

Strand Map

Benchmark

15 Research

See how the benchmark is part of the coherent story told in *Science for All Americans*.

SFAA

Third, the group should continue with new topics, gradually building a record of what topics and goals will be part of the curriculum at each grade level and what changes have to be made in the curriculum to accommodate the decisions. As this process continues, the participating faculty members will be able to increase their knowledge of student learning goals and hence help to raise the capability of the faculty as a whole to engage in curriculum reform.

Conducting Benchmark Workshops

The CD-ROM *Resources for Science Literacy: Professional Development* describes workshops designed to help educators understand *Benchmarks* (and, by virtue of their similarity, *NSES* and NCTM's *Standards* as well). Its many embedded options enable a person or group to create a workshop of almost any duration tailored to the particular interests, needs, and circumstances of a group of educators. As the summary of the CD-ROM's workshop component opposite suggests, workshop designers have access to outlines, scenarios, handouts, transparency masters, references, and sample presentation scripts to produce a wide variety of workshops for their colleagues.

Becoming Familiar with Research on Learning

Teaching is more craft than science. Over the years, teaching has been shaped informally by what seems to work best in the individual teacher's classroom, by broad theories about student learning, and by tradition. But for the most part, practices based on those experiences, theories, and traditions have not been put to rigorous tests to see if they really are sound. Similarly, teachers and other educators (textbook developers, framework writers, test makers) necessarily have had to base curriculum decisions on shared beliefs that have accumulated over the years on what students can learn, beliefs that until recent years have rarely been examined systematically.

Gradually, however, research is identifying some principles of teaching and learning that apply rather generally and some that apply to specific content areas. For example, giving a student enough time to formulate an answer thoughtfully (the well-known "wait time") may be a universal principle of good teaching. Much more specifically, quite a bit is known about when and how students seem to be able to learn about (and believe in) the molecular model of matter, the shape of the earth, the distinction between heat and temperature, how plants make their food, natural selection, and the equivalence of fractions and decimals. Currently, there are still large numbers of concepts for which little learning research has been carried out. Perhaps the focus offered by the reduced and highly specific ideas in national goals will stimulate interest in—and funding to support—a greatly intensified research effort on how to promote learning of those ideas.

There are many accounts of creative teaching ideas to be found in teacher magazines and journals, which usually focus on whether lessons "work" in the classroom. This literature can be very stimulating, but seldom offers convincing evidence for what particular ideas or skills students may have learned. The advent of specific national learning goals may help to focus that literature, too, on demonstrating results.

DESIGNING PROJECT 2061 PROFESSIONAL DEVELOPMENT WORKSHOPS

The Workshop Guide, found in the Project 2061 CD-ROM *Resources for Science Literacy: Professional Development*, provides advice, example scripts, and materials for designing a variety of workshops—including a large store of transparencies, handouts, and readings.

All Project 2061 workshops aim to demonstrate the need for reform and then a particular way in which Project 2061 tools can help (for example, analyzing curriculum frameworks). Workshops usually have three major stages: (1) general reform rationale, (2) particular use of tools, and (3) reflection and summary. Within each stage, two to ten options are provided for each of several steps. After first deciding which tool use to focus on, workshop designers select other options that will best suit it, meet the participants' needs, and fit the time available. To help prospective workshop leaders get started, three examples of complete workshop agendas are also provided for six-hour, one-day, and two-day workshops. The Workshop Guide utility itself can also serve as a tutorial for learning more about Project 2061 and how to use its tools for science literacy.

This is the general workshop format as presented in *Resources for Science Literacy*:

Opening
This stage allows the workshop leader to find out more about what participants already know about science education reform and Project 2061 and to establish the specific learning goals for the workshop.
- Introduction (four options available)
- Need for Change in Science Education (six options available)
- Workshop Goals (four options available)
- What Do Participants Know about Project 2061? (three options available)

Project 2061 Tools
This is the core of a Project 2061 workshop, and the option chosen here will help determine which Opening and Closing options are selected.
- Overview of Tools Available from Project 2061 (ten options available)
- Exploring the Use of Project 2061 Tools to (choose one of the following uses):
  - Understand the nature of benchmarks (five options available)
  - Analyze curriculum frameworks (one option available)
  - Analyze curriculum materials (two options available)
  - Analyze instruction (five options available)
  - Design instruction (seven options available)

Closing
In this stage workshop participants reflect on what they have learned and provide the workshop leader with feedback on the effectiveness of the workshop itself.
- Summary (five options available)
- Evaluation (six options available)

As the body of systematic knowledge about teaching and learning grows, educators can turn to it more frequently than in the past for guidance in making informed curriculum decisions. This does not mean, however, that all educators in a school district need to be trained as researchers, or even as skilled analyzers of research. The professional capacity of a school system can be enhanced if some of the faculty accept responsibility for keeping up to date on what the research (see marginal note) says about learning and teaching and for locating pertinent research information when needed. Several different teams should be formed to track research for the school district, since most cognitive research on learning focuses on the learning of particular concepts of a particular subject at a particular grade level. Distribution of interesting findings to faculty in the relevant areas would, of course, be part of this task.

A school-district group concerned with research on learning in science, mathematics, and technology could start by becoming acquainted with the scope of cognitive research in those fields and with what implications it has for practice. From time to time, members of the group will want to read and discuss some of the original research accounts to see precisely what was done and found out. But most often their task will be to relate research findings to curricular issues.

From *Benchmarks* CHAPTER 15: THE RESEARCH BASE, the group could turn to the more elaborate, annotated references to the research literature and to the summary accounts of research provided in *Resources for Science Literacy: Professional Development*, which are also keyed to sections of *Science for All Americans* and *Benchmarks*. Other helpful places to look for findings from research are the *Handbook of Research on Science Teaching and Learning* (Gabel, 1994) and the corresponding *Handbook of Research on Mathematics Teaching and Learning* (Grouws, 1992); the series of volumes entitled *What Research Says to the Science Teacher* and the mathematics collection *Research Ideas for the Classroom* (Jensen, 1993; Owens, 1993; Wilson, 1993) with volumes for three different grade ranges; and the special mathematics, science, and technology chapters in *Handbook of Research on Curriculum* (Jackson, 1992).

LEARNING TO ANALYZE CURRICULUM MATERIALS

Educators select instructional materials to serve a curriculum, or perhaps it is the other way around—materials that educators select determine the curriculum. Either way, making decisions about instructional materials is a major professional responsi-

"Even after some years of physics instruction, students do not distinguish well between heat and temperature when they explain thermal phenomena (Kesidou & Duit, 1993; Tiberghien, 1983; Wiser, 1988). Their belief that temperature is the measure of heat is particularly resistant to change. Long-term teaching interventions are required for upper middle-school students to start differentiating between heat and temperature (Linn & Songer, 1991)."
—*Benchmarks for Science Literacy*, p. 337

bility in every school district, even in ordinary circumstances. When a district is engaged in curriculum reform, the evaluation of instructional materials takes on still greater importance. Building a professional capability for reform therefore includes making sure that those individuals who will make decisions about the selection of curriculum materials acquire the technical knowledge and skills for analyzing materials and comparing their advantages and disadvantages. Moreover, teachers who are not themselves directly engaged in the process should know enough about it to be able to interpret and respond to the recommendations.

Asking whether curriculum materials would actually help students to achieve benchmarks is a powerful way to understand both better. In the past, the evaluation of curriculum materials has often been sporadic (taking place for any subject only once every five years or so), free floating (paying little or no attention to articulation across the grades), ad hoc (being conducted each time by a new committee of mostly novice evaluators), and short (having to be accomplished usually in a matter of hours or days on top of regular assignments). The evaluation process typically has been highly subjective, with little basis for estimating whether the conclusions reached are consistent ("reliable") and true ("valid").

Shortcomings of time and effort are matters of administrative priority and can be easily changed once the task of curriculum-materials analysis is taken seriously enough. More difficult to correct is the absence of two resources: (1) a set of coherent and authoritative specific learning goals as the chief criterion for judging curriculum materials, and (2) a rigorous analytical procedure for examining curriculum materials in the light of their likely contribution to the achievement of those learning goals.

As energetic and conscientious as the efforts of some educators have been to design sound curriculum without these resources, a new level of effectiveness is now widely possible—at least in science, mathematics, and technology education. National benchmarks and standards provide the needed learning goals at a suitable level of specificity. And even as Project 2061 was developing *Benchmarks*, it began exploring how specified learning goals could actually be used effectively to help make decisions about such practical matters as teaching, testing, and curriculum materials. A system of analysis has been developed by Project 2061 that focuses on the attainment of specific learning goals and describes the kind of evidence required to make a case for what students are likely to learn. See the following pages for a description of the system and the criteria it uses.

Judgments by independent reviewers are said technically to be "reliable" if they are similar to one another; whether they are true is another matter. Reviewers may share a bias or misunderstanding, and hence make consistent but false judgments.

ANALYZING A CURRICULUM MATERIAL

Step 1: Identify likely benchmarks. Make a list of a few benchmarks that are important and that you would expect the material to focus on. Next, look through the material to find instructional experiences that might help students learn those benchmark ideas. If you can't find such evidence for a particular benchmark, then cross it off your list. That will give you a much shorter list of benchmarks on which the material actually focuses.

Step 2: Clarify the benchmark's meaning. Pick one benchmark from the short list and study it, as described on pages 188 and 189.

Step 3: Reconsider how explicitly the material targets the benchmark. Now go back and briefly describe the evidence and where you found it in the material, including the teacher guide. Consider whether the activities are appropriate for the intended grade level. You may find, for example, that a set of lessons on the water cycle, although advertised as K-2, focuses on evaporation and hence targets benchmarks for grades 3-5. For activities that are appropriate, is there adequate guidance to ensure that the benchmark idea will be addressed?

Step 4: Estimate how effective the instruction would be. Now use the best available knowledge of how students learn to reflect on how much students would actually learn about the benchmark from the recommended instruction. Project 2061 has developed criteria for estimating the effectiveness of instruction. (See facing page.) For example, do the recommended activities provide students with *memorable* experiences, opportunities to *reflect* on them, and opportunities to explore concepts in *varied contexts*? Even good teachers find that they often underestimate how difficult some ideas are for many students, especially when the students already have persistent misconceptions.

Step 5: Summarize and make recommendations. Review your findings on all the criteria and summarize how effective you think the material, together with what *appears explicitly* in its teacher guide, would be for helping students to achieve the benchmarks. Consider what it would take to improve it—for example, increasing the variety of phenomena studied, providing better questions to guide students' thinking, and making the assessment fit the specific benchmarks better. Finally, recommend how the material should (or should not) be used in the curriculum.

CRITERIA FOR ESTIMATING INSTRUCTIONAL EFFECTIVENESS

Project 2061's curriculum-materials analysis procedure uses the following questions to determine the extent to which a material's instructional strategy is likely to help students learn the content. Each question focuses on *specific benchmarks*, not just content in general. To what extent does the material:

Provide a sense of purpose? That is, does it
... provide an overall sense of direction that students will understand and find motivating?
... provide a sense of purpose for each lesson and its relationship to others?
... provide an obvious rationale for the sequence of activities (versus just a collection)?

Take account of student ideas? That is, does it
...specify prerequisite knowledge/skills?
...alert teachers to commonly held student misconceptions?
...suggest how to find out what students think about relevant phenomena and principles?
...explicitly address commonly held student ideas?
...include suggestions for teachers on how to address ideas that their students hold?

Engage students with phenomena? That is, does it
...include direct experiences (or close approximations) with relevant phenomena?
...promote experiences in multiple, different contexts to foster generalizations?

Develop and use scientific ideas? That is, does it
...build a case for scientific ideas based on their success in explaining phenomena?
...link technical terms to experiences and only when needed for communication?
...include a variety of representations of scientific ideas that are both comprehensible and valid?
...tie ideas together over time logically and explicitly?
...explicitly draw attention to appropriate connections among benchmark ideas in different topics?
...describe how teachers can demonstrate application of skills or knowledge?
...provide tasks on which students can practice application in a variety of situations?

Promote student thinking? That is, does it
...encourage each student to express, clarify, and justify—and get feedback on—his or her ideas?
...include sequenced tasks or questions to guide student reasoning?
...help or suggest how to help students to know when to use knowledge and skills in new situations?
...suggest how students can check their progress and consider how their ideas have changed?

Assess progress? That is, does it
...include tasks to assess student achievement of particular benchmarks?
...include tasks requiring new application, not just plugging into formulas or repeating definitions?
...embed assessment in instruction with advice on using the results to choose or modify activities?

Promote other benefits? That is, does it
...improve teachers' own understanding of science, mathematics, and technology and their connections?
...foster student curiosity, creativity, and healthy questioning?
...foster high expectations and success for all students?
...explicitly draw attention to appropriate connections to ideas in other units?
...have other features worth noting?

Even when highly able teachers are carrying out the evaluation procedure, it is good practice for them to follow detailed steps aimed at counteracting old short-cut habits that were acquired under typical conditions of inadequate study time. Separate, summary judgments about content match and instructional quality are often seen not to hold up when the particular instruction for particular ideas is investigated. The key question has to be, "How likely is it that all students will learn this particular idea from these prescribed activities?" There is no question that the analysis procedure is demanding of time and effort, but experience with a variety of simpler alternatives shows that reliable and valid judgments require such investment. It is possible however, that experienced analysts will eventually be able to shorten the procedure.

The Project 2061 procedure for analyzing curriculum materials needs to be conducted by teams that already have expertise in content and instruction and are able to use the analysis criteria accurately and reliably—which requires intensive training and practice. That is a tall order, and many school districts may well find it too daunting to undertake. For that reason, Project 2061 is in the process of training national teams to evaluate curriculum materials and is slowly building an on-line database of science, mathematics, and technology materials that have been evaluated by those teams. The teams' evaluation reports will not make yes-or-no recommendations but rather present profiles showing how well the materials appear to support student progress toward benchmarks (or other coherent sets of specific learning goals). But even when such analyses become available, they are not likely to be used well unless users understand enough of the process involved to make sense of the profiles and to determine their implications for making decisions. *Resources for Science Literacy: Curriculum Materials Evaluation* provides guidance for developing both general familiarity with the process and the technical skills required for its implementation.

ACQUIRING CURRICULUM VERSATILITY

Project 2061 does not advocate any particular kind of instructional format, scheduling of time, or form of assessment. Nor does it endorse any particular organization of instruction—thematic or academic, integrated or disciplinary, cooperative-group or competitive, lecture or hands-on. Different content and learning goals are likely to be better served by some formats than by others. Moreover, a wide variety of research findings suggest that different students learn different ideas in different ways. But the evidence does not yet provide a sound basis for predicting the best way for a particular student to learn a particular idea from a particular teacher. Perusal of the literature on trials of seminars, design projects, independent study, etc. show that these have seldom had the benefit of well-defined goals and well-controlled circumstances. Again, the improved focus afforded by national goals may expand and improve the knowledge base. We can be fairly sure that no single approach to the organization of instruction will be found to be consistently best for all students. So in anticipation of documented advantages and disadvantages of different instructional strategies, it is a good idea to develop some collective experience with a variety of formats.

Future developments in curriculum design and in instructional technology may make possible many more options for organizing instruction than are now available. High-quality curriculum blocks developed in the future may require a variety of different instructional strategies. For a school district to consider those possibilities seriously, it needs to appreciate their individual advantages and limitations and know what it takes to operate them well with local students. Project 2061 is not pressing for maximizing curriculum variety as an end in itself, but for freeing districts to use the latest resources available, regardless of their format. If none of the faculty in a school district has firsthand experience with other than the traditional way of operating a curriculum, then there will be no basis for making informed decisions when alternatives are proposed. To develop the needed professional capability, a school district should identify groups of teachers who will, through study and experience, become knowledgeable in the conduct of different instructional formats and time patterns. In addition, there ought to be teachers at every grade level who have used the various formats in actual instruction.

Alternative Instructional Formats

No one teacher needs to be expert in the conduct of every possible format, but collectively a school district should have teachers with experience in the use of a variety of formats. In this way, a district will have a base for making informed decisions on the adoption of different formats—and to be in a strong position to introduce new formats if creditable curriculum blocks require them.

A convenient way to get started building such a capability is to set up a conference "listserv" on the district computer network for each of the formats of interest at each

Although there is only sparse research on how well alternative formats work, *Designs on Disk* includes a variety of articles about developing and using them.

The *Designs on Disk* database for reporting on and discussing instructional formats offers a convenient source of ideas for school districts to test on a limited scale.

grade level. Thus, there could be elementary-school, middle-school, and high-school conferences for teachers and administrators interested in seminar, project, independent-study, peer-teaching, integrated-course, and discipline-based course formats. The members of each conference would share in building an electronic library of outstanding articles having to do with the theory and application of its format. A next step would be for some teachers to implement one or more of the proposed ideas. That would entail describing the process, keeping notes on what transpired, and sharing the results with the conference members.

Consider the following. An elementary-school teacher may decide to try both independent study and peer teaching with regard to basic arithmetic skills. After students have demonstrated that they can perform certain specified paper-and-pencil math calculations, they are notified that they must master related (and also specified) handheld-calculator skills before such and such a date—but that no class time will be used for that purpose. They are on their own, but if they wish, they can get help outside of class time at a computer clinic staffed entirely by (selected and trained) students. The teacher keeps notes on how students respond to the assignment and what the learning results are, and summarizes the episode for conference colleagues. (It would be necessary, of course, to keep an eye out for deleterious side effects that would compromise the success of some students. And it would be a good idea to be sure parents approve of the experiment.)

It is well to remember that a good alternative may not work terribly well the first time or two it is tried. Students as well as teachers need some time to get used to new procedures and expectations. Advice from experienced practitioners is highly desirable, for there are usually requisites that are not mentioned in the promotional literature. It would be helpful if enthusiasts who write articles about novel instructional formats disclose the difficulties as well as the triumphs (as, for example, advocates of cooperative groups might describe frankly the effort required to shape and sustain cooperation). Without this understanding that new approaches almost always take time to perfect, many good ideas may be discarded prematurely.

As teachers in different grades try different formats with different content and in various situations, they may elect to adopt them as a more or less fixed part of their program, retain them to be used some years but not every year, or decline to make further use of them. As their efforts are documented and entered in the district's curriculum ideas file, other teachers, particularly those participating in the relevant network conference, will be encouraged to duplicate them, with the result that there

Traditional formats have their own side effects, inequities, and failures—but they are at least familiar and we more or less have learned to live with them.

may be a gradual increase in the use of seminars, projects, independent study, and peer teaching in the district. One sure result will be to increase the professional capability of the faculty.

Alternative Time Patterns

Much the same argument can be made for setting up computer-network conferences of teachers and administrators to explore various scheduling arrangements, with the objective of finding ways to pursue specific learning goals more successfully. Again, the purpose is to have some faculty members at every grade level and in every subject matter become knowledgeable about the advantages, limitations, costs, and risks associated with variable time scheduling that could accommodate (or be required within) highly valued curriculum blocks. Conference participants can build a new database of ideas for flexible time scheduling, and then teachers can volunteer to try out some of the ideas and document the experience.

Trying out alternative time arrangements may be more difficult to carry off than trying out alternative format ideas, but not prohibitively so, especially if several teachers take it on together. Some informal experimentation ought to be possible in most school situations. The basic rationale should be that a desirable unit or course requires a different time frame to realize its full effect. If the rationale seems cogent and results appear to be promising, then approval should be sought for trial on a larger scale, engaging administrators, parents, and students. Having to get approval gives another reason, in both cases, for spelling out in writing the nature and limits of the trial, the rationale for conducting the trial, and what information will be collected on its operation and effects.

Some kinds of alternative time arrangements are relatively easy to test. In the elementary grades, self-contained classrooms are the norm. In principle, a teacher can configure daily instruction time in many different ways as long as the amount devoted to each subject over a longer time span—say, a semester—meets district and state requirements. "Block scheduling" on alternative days or semesters is becoming increasingly popular, and "modular scheduling" that allows for a variety of possible period lengths crops up every now and then.

Suppose, for example, that an average of 20 minutes per day of science is required. A teacher may consider apportioning that as two consecutive periods once a week, as 10 minutes a day plus a period once a week, or as nearly all day once every other week. The 10-minute-a-day alternative would best suit an

Some of the conveniences of standard schedules are mentioned in the Chapter 2 section on schedule variations.

Period	Monday	Tuesday	Wednesday	Thursday	Friday
1	Geometry	Geometry	Geometry	Geometry	Geometry
2	English	English	English	English	English
3	Spanish	Spanish	Spanish	Spanish	Spanish
4	PE	Semester Project	PE	Semester Project	PE
Lunch	Lunch	Lunch	Lunch	Lunch	Lunch
5	Music	Music	Music	Music	Music
6	Global Studies	Biology	Global Studies	Biology	Global Studies
7	Biology	Global Studies	Biology	Global Studies	Biology
8	Biology	Global Studies	Biology	Global Studies	Biology

extended series of daily measurements; the all-day alternative would best suit an experimental problem that requires a long setup time or iterative trials. Such arrangements permit students to engage in both extended hands-on investigations and discussions of their findings, and they make it possible for elementary teachers who specialize in science to be used effectively. The same can be said for mathematics, art, and some other fields.

In middle and high schools, instruction is usually departmentalized, but that does not altogether preclude gaining experience with various nontraditional time arrangements. Following are two quite different possibilities, the first a fairly common experiment, the second very rare.

In the first instance, arrangements are made for paired classes in which the same students are enrolled in two courses that meet in successive periods. (Science and mathematics have been thought to make a good combination, as do, for example, mathematics and physical education, science and technology, science and history, mathematics and social studies.) The two teachers then have ten periods a week to use in various ways, such as giving each student one double period and three single periods a week in each subject. This setup offers the possibility of extended instruction and of intersubject collaboration.

In the second instance, arrangements are made to commit one period each week to a genuine seminar (not "class discussion") on a trade book. This takes place over a whole quarter or semester of a conventional one-period-a-day science course. To keep

A variety of ways of partitioning time are described in Chapter 3.

the seminar a reasonable size, the class is divided into three groups that explore different trade books on the same topic. The assumption is that such a seminar could enhance achievement of benchmarks in certain chapters of *Benchmarks for Science Literacy* (THE NATURE OF SCIENCE, THE NATURE OF MATHEMATICS, THE NATURE OF TECHNOLOGY, HISTORICAL PERSPECTIVES, HABITS OF MIND, and COMMON THEMES) that deal with ideas that may not easily be learned through more closely managed instruction. But if the seminar format is initially more time consuming, some marginal topics could be eliminated from the course as a whole to accommodate the reduced amount of time available for conventional instruction.

In all such trials, keeping notes on the process and sharing results with colleagues should be part of the plan. Because the climate should be experimental, with the expectation that not everything will go smoothly the first time, there is no need for accounts to be as sunny and unblemished as in most journal articles. Difficulties and shortcomings can and should be reported, along with successes.

The *Designs on Disk* database of instructional formats can be of help in preparing reports and sharing them.

IMPROVING ASSESSMENT

It is essential to know something about how well each student is doing individually. Good teachers assess student performance frequently because effective teaching depends on having accurate and detailed information on what students do or do not understand. On a more summary level, parents need to know how well their children are doing in order to help fulfill their responsibilities properly; and eventually potential employers and admissions officers will base decisions about each student on his or her performance record. It is also important to estimate how effective the educational program is collectively. Schools, school districts, states, and the federal government periodically need to determine how well their populations—and defined subpopulations—of students are performing so they can make informed policy decisions about instructional practices and curriculum choices.

These three needs—to monitor the day-to-day learning of individual students, to summarize their individual progress and achievement, and to monitor the effectiveness of an educational program for the progress of defined populations of students—are often confounded. Judgments about individuals are ideally based on many sources of information, including what can be gleaned from quizzes and tests, homework assignments, essays and projects, portfolios of student work, teacher observation, and interviews. To report to third parties on the progress of students doesn't require nearly the amount of detail that the teachers themselves need. A single letter grade is usually too simple, greatly underrepresenting the very different possible profiles of knowledge and skills underlying it. But a summary profile of achievement, perhaps supplemented by brief commentary, may be satisfactory for many purposes (and can be followed up in more detail if desired).

To ensure fairness in judging individual students, all students should probably be subjected to the same assessment tools, or at least sets of equally difficult and representative tasks. But typically, individual tests cannot tap anywhere near all of what one would hope students have learned, but must settle for being a good sampling of that domain. And typically, the particular sample to be tested is unknown to students in advance, to prevent their study being narrowed to just that part of the domain.

To find out what a population of students has learned requires a very different approach. The simple, popular notion of giving the same test to every student in the population is monumentally wasteful. Acquiring complete and identical information on every student is far too expensive in terms of use of students' time and collection costs. However, reliable estimates about large groups can be made by sampling relatively small numbers of students—which, paradoxically, makes possible much better information. Rather than give the same test to 10,000 students, ten different tests covering ten different areas can be given to 1,000 students each. The sample size of 1,000 is enough to ensure a reliable estimate of what all 10,000 know on each test, and a ten times greater area of learning can be probed (or the same areas in ten times more detail) for the same investment of time. Moreover, a greater variety of performance can be assessed. In the same amount of time (say a one-hour examination), some students could get dozens of short-answer questions, others a dozen problems, and others a single higher-order problem-solving task.

Of course, results on such a large scale are not useful for reporting on individual students, because only a very small part of any one student's possible performance is tapped, and the different tests are not likely to be (nor need to be) precisely equivalent in difficulty. And, although such large-scale testing can produce far more information about the performance of a population of students, there is a new source of uncertainty: because results cannot be used to judge individual students, the motivation of students to show how well they can do would seem likely to be less. Because certain categories of students may experience a greater drop in motivation than others, not only the absolute but even the relative achievements may be misestimated. (Assessment experts are still working on the motivation question.)

Alignment of Assessment and Curriculum

There are few topics in education that create more controversy than assessment does. In addition to disagreements over what techniques and instruments to use, there are differences over what the alignment should be among assessment, curriculum, and learning goals. The ideal, of course, is that curriculum and assessment be aligned both with each other and with specific learning goals. In a context of rapid change, however, there is inevitably some jostling about where change should begin. Commonly, the discussion revolves around two views—what may be called the "fairness view" versus the "leverage view."

The fairness view requires that assessment of students be aligned with the current curriculum. Students, teachers, and parents understandably want assessment to be fair, and fairness demands that students not be judged on what they have had no opportunity to learn. In this view, therefore, assessment should be aligned with the current curriculum—whatever it is. If and when the curriculum is changed, the assessment should change correspondingly. Thus, curriculum reform should lead and assessment should follow.

On the other hand, the leverage view requires that the curriculum be aligned with a new assessment. Many educators and citizens see assessment as a strategic lever for improving the curriculum. The proposition is that teachers will make (and parents and policymakers will welcome) whatever changes in the curriculum are necessary to help students do well on the tests. Change can more easily be made in assessment than in curriculum, and curriculum can be expected to adjust to assessment naturally. Thus, assessment reform should lead and curriculum reform will follow.

Both of those positions, obviously, imply a prior alignment. If assessment is to be aligned with the curriculum, what is curriculum to be based on? Or if assessment is to take the lead, from where does it obtain its authority? Curriculum and assessment are both means to ends, not ends in themselves. Hence, there is a third point of view that argues that curriculum and assessment both ought to be based on the same set of established learning goals. In this view, curriculum and assessment are independently based on specific learning goals, and neither need be explicitly adjusted to the other. Schematically, these three points of view are parts A, B, and C in the box that follows.

Realistically, the completely independent approach is not likely to be satisfactory. One possible reason is that neither the curriculum nor the assessment will represent all the learning goals equally. If the balance arrived at in the curriculum is different from the balance arrived at in the assessment, there could be a significant mismatch.

CURRICULUM AND ASSESSMENT RELATIONSHIPS

A. Assessment follows curriculum

An alternative to the assessment-first strategy of curriculum change is to begin with curriculum—which presumably has already been aligned with goals—and then align the assessment to what is actually taught. A danger here, however, is that the assessment will be tailored to fit incidental aspects of the curriculum that go outside of what was specified by the goals.

B. Assessment leads curriculum

Concern about the difficulty of changing curriculum leads to calling for changes in assessment as a means of inducing teachers to change what they teach. This instrumental use of assessment may arise from a passion about assessment methods per se, or from a concern about propagating a set of goals—to which the assessment itself would have to be already aligned.

C. Curriculum and assessment both derived from goals, independently

Another view is that curriculum and assessment should both be aligned to the same specific learning goals, but by different groups. Differences in perceptions or preferences of the groups may produce uncomfortable discrepancies. Misalignments could also occur if the two groups focused on different subsets of the common goals.

D. Curriculum and assessment both derived from goals, in parallel

Congruence between curriculum and assessment can be improved by settling on specific goal emphasis for both in advance. Or, at least, drafts of curriculum and assessment could be reconciled to each other.

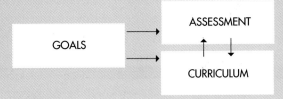

A second reason for not deriving curriculum and assessment independently is educators' uneasiness about making assessment of learning goals completely free of the particular context in which the goals were learned. Consider, for example, a benchmark on measurement skills in which the intention is that students could think about errors of measurement generally, in any context in which they are making measurements. But if measurement skills are learned largely in the context of studying oxygen levels in a local lake, teachers of that curriculum would be likely to assess student skill in those very measurements rather than, say, in a novel context of weather measurement. (Research does show, it is true, that learning does not easily transfer to new contexts.)

So a more practical plan would be to attempt to derive both the curriculum and the assessment from learning goals, but with each keeping an eye on the other and making accommodations from which each is likely to benefit. Regarding measurement skills, for example, the curriculum designers could build in more variety of measurement contexts to promote better transfer; and the assessment designers could provide students with a chance to show that at least they can measure in familiar contexts. This parallel alignment approach is illustrated by part D in the box opposite.

If a school district were starting out with a clean slate, the order of alignment would matter little in principle as long as ends (learning goals in this context) came first. In reality, though, few districts start with a clean slate: they already have a curriculum, though they may be in the process of changing it. On the other hand, few districts have a systematic process for assessing the effects of their curriculum. Thus the suggestions for action that follow concentrate on beginning to develop a school or school-district capability for curriculum evaluation that is designed to give good estimates of how well students in the district are reaching specified learning goals.

Experimenting with Assessment

One can imagine a time when each course, large unit, or other block in a curriculum contains built-in assessment materials and procedures that have been carefully validated so that the block produces the student learning that the assessment measures. A district could then ascertain whether students are reaching the specified learning goals block by block (or at least at the grade-range checkpoints), and the sum of its findings would indicate what the curriculum as a whole has achieved. But good curriculum blocks appear to be more rare and more demanding to construct than may have been thought, so an adequate pool of blocks may be a long time in coming. In any

Examples of assessment alternatives for a single benchmark on sampling

Multiple choice (30 seconds):

The most important feature of a scientific comparison of two groups is

    a. how group members were selected

    b. the size of the groups

    c. what percentage of the whole population do the groups represent

    d. what kind of average is used to represent the groups

Constructed response (5 minutes):

Kim and Keisha had an argument over whether an MTV video was more popular among boys or among girls. Describe how they could choose groups of boys and girls to ask to find out who was right.

Problem solving (5 days):

Find out whether an MTV video is more popular among boys or among girls in your school.

case, it is possible to monitor a curriculum without having to monitor every course separately and completely.

It is important to distinguish the purposes and strategies of monitoring students as individuals from those of monitoring the effectiveness of the curriculum. Although the pool of tasks used for assessment may be the same, the tasks may be selected and employed in significantly different ways. For example, a much larger set of tasks can be used for curriculum monitoring by giving different students different subsets of tasks. The effectiveness of the curriculum can be estimated satisfactorily from the partial results and, since students need not be graded on the tasks, the different subsets need not be of precisely equal difficulty. The scoring need not be as elaborate for curriculum monitoring, since the number of students passing some threshold of success on each task may be all the information needed. For student monitoring, on the other hand, tasks can be selected more narrowly to fit the students' recent learning experiences and ability, yet interpreted more elaborately to diagnose the individuals' levels of understanding.

Monitoring Classroom Learning

Curricula of the future are likely to demand that teachers are able to use a variety of conventional and nonconventional techniques for probing student learning. Skill in using multiple-choice tests, essays, portfolio analysis, oral interviews, and other approaches to assessment depends on understanding their advantages and limits, knowing when to use them, and being able to interpret their results correctly.

Although a high-quality assessment program requires long and reiterated development, a good way for a school district to develop its teachers' capability to employ such programs eventually is to have groups of teachers actually create samples of each kind of assessment and use them with students. The basic sampling process consists of the following five steps:

First, each group (subject and grade related), should select a single, specific learning goal of interest to the group and analyze it carefully to reach agreement on what it really means—and does not mean. If the learning goal is too general, the process may go off on different tangents. (Techniques for analysis are described earlier in this chapter.) The strand maps in *Atlas of Science Literacy* can be consulted in identifying precursors and connections to the key idea.

Second, subgroups of two or three teachers each should develop a different way of determining whether students have acquired the knowledge or skill defined by the selected benchmark (say, a set of objective short-answer items, an essay question, and a student interview). Research articles on student misconceptions may provide a starting point for identifying appropriate questions.

Third, at a suitable time, the subgroup teachers should apply their method to some students and use the responses to estimate the degree to which the learning goal has been reached by individual students. Each subgroup should write up the episode as a brief case study.

Fourth, the teachers should discuss their experiences and findings. The discussion should include how well the judges agree on the scoring and how well the assessment gets to the intended understanding or skill. Since discussion of the scoring can often lead to argument about just what each goal really means, the activity should increase teachers' understanding of the goals and their sensitivity to what constitutes student understanding.

Fifth, after discussion of the case studies, the group should select another benchmark—perhaps from those that strand maps show are closely related to the first one, or perhaps from a different content area entirely. For the new benchmark, each subgroup could change to a different assessment approach. Two or three cycles should be ample for all participants to become familiar enough with assessment techniques to make informed use of them immediately, if they wish, and to be prepared to employ them in redesigned curricula in the future.

Although some teachers may be inspired to develop more assessment tasks, others, who recognize how hard it is to invent good tasks, will settle for being better choosers and interpreters of tasks developed by specialists.

Monitoring Curricula

School districts that set up systems for monitoring the effectiveness of the curriculum usually conduct assessments of student learning at a few specified checkpoints—for example, at the end of grades 4, 8, and 11. For districts guided by *Benchmarks for Science Literacy*, the checkpoints of choice would be at the end of grades 2, 5, 8, and 12, and committees of teachers would be set up to develop some prototype assessments in science, mathematics, and technology for those grades. In the first year, a relatively few benchmarks would be selected and tests developed for them. At the appropriate time, a sample of students from all of the participating teachers would be tested, the data aggregated (so as not to be teacher specific) and then analyzed and discussed by the committee. In the second year, a larger selection of benchmarks would be made, and the process repeated. In this way, a district can gradually build a method for finding out how well its students are progressing toward science literacy, and, in the process, it can significantly increase the professional capabilities of its staff.

School districts usually have some staff who understand the design and statistics of large-scale assessments of learning and who can be called upon to provide leadership in building a capability for curriculum monitoring.

BECOMING INFORMED ON REFORM MOVEMENTS

Finally, each school district should know what is going on nationally and statewide with regard to K-12 curriculum reform so that it can benefit from the concentration of resources and skills that are possible in large-scale projects. These projects issue reports that follow up on earlier reports and increasingly share their ideas and products more promptly and often on the Internet. Not every teacher needs to be so well informed. Nor is this to say that a district should feel compelled to introduce every new reform—or claimed reform—that surfaces. Rather, it should put itself in a position to decide knowledgeably which reforms to reject outright and which to consider further—maybe even which to adopt.

Committees should be set up in districts to track developments in a particular sector defined by subject matter, grade level, or aspect. Members would bring interesting ideas to the attention of colleagues (teachers, administrators, and school-board members), and serve as internal consultants on curriculum-reform matters. Nonetheless, some of the curriculum committees should cut across grade levels and subject-matter fields, because many ideas for reform follow that pattern and because it helps to moderate the tendency of specialty groups to take a narrow view of the curriculum. Of course, the ability of these committees to be of real service depends on their having ready access to reports, newsletters, journals, books, and the Internet, along with office support and a budget for travel to conferences and to schools implementing reform of interest. In other words, building a professional capability for reform in a school district calls for support of work that includes but is not limited to classroom teaching.

The suggestions in this chapter, helpful as they may be, all imply more demand on teachers' time—a commodity already in painfully short supply. In the long run, changes in policy and financing are needed to make more teacher time available for planning and professional development. Yet, while there is no easy solution to the inadequacy of time for teachers' planning and professional development, such time as there is could be used more efficiently if it were better focused. In science, mathematics, and technology, that focus should be on promoting student learning of the carefully selected and coherent set of core ideas on which there is currently a wide national consensus. Project 2061's experience is that many dimensions of professional devel-

Project 2061's Professional Development Programs offer educators a variety of workshops and other services that can be customized to fit the long-term needs of each school or district.

opment can benefit from such a focus, including improved knowledge of subject matter and what aspects of it are most appropriate for students, better understanding of the general psychology of teaching and learning, skills in judging materials for instruction and assessment, and the value of collaboration with peers.

In the next chapter, consideration is given to how to reduce the number of topics taught. Abandoning less important topics will not directly make more time available for teachers, because the time that is freed up should be used to teach better the central ideas that remain. The tighter focus that will result from this effort will enable educators to make better use of the precious planning and professional-development time that is currently available.

"Basically, we're all trying to say the same thing."

B. Kliban, *Cat*, 1987

CHAPTER 7
UNBURDENING THE CURRICULUM

Time in school for teaching and learning is not limitless. Yet many textbooks and course syllabi seem to assume otherwise. They include a great abundance of topics, many of which are treated in superficial detail and employ technical language that far exceeds most students' understanding. And even as new content is added to the curriculum—little is ever subtracted—students are being asked to learn with greater depth. Rarely is more time made available for accomplishing this. Coverage almost always wins out over student understanding, quantity takes precedence over quality.

Many decades of overload have shaped curriculum, textbooks, tests, and teacher expectations into an industry of superficiality. Many teachers know, or at least suspect, how little their students understand, but do not know how to transform the system. Lengthening the school day and year and reducing the number of different subjects students study are obvious though apparently unpopular remedies for the mismatch between curricular time and content, but in any event would not by themselves solve the problem.

Another remedy sometimes proposed is to move concepts lower in the grade sequence, thereby leaving time free in high school for students to learn better what they study. That would be a tenable ambition if high-school students were learning those ideas well now. Because they are not, however, it is no more than wishful thinking to believe that younger children will be able to learn what older children apparently do not.

Improvements in teaching methods and curriculum design may eventually make it possible for students to learn more than they do now, hour for hour, but the current and critical need is for them to acquire at least some important knowledge and skills better, even at the price of covering fewer topics overall. This chapter describes four strategies aimed at reallocating time—time to focus on understanding important facts,

Some researchers in science education estimate that even good students understand and retain only a small fraction of what they study. As disturbing as this claim is, there is a great deal of research—on school children, college students, and adult citizens—that substantiates it. See *Benchmarks* CHAPTER 15: THE RESEARCH BASE.

The pernicious effects of an over-stuffed curriculum is one of the major messages in the Third International Mathematics and Science Study (TIMSS) reports of 1997-1998.

principles, and applications in science, mathematics, and technology, not time to enable still more material to be superficially covered. The underlying purpose is to realize a better cost-to-benefit ratio, using time and resources in ways that will maximize students' eventual science literacy. The strategies are:

- Reduce the number of major topics taught.
- Prune some topics by removing unnecessary details.
- Limit technical vocabulary to essential terms.
- Eliminate wasteful repetition.

CUTTING MAJOR TOPICS

Although the ambiguous meaning of "topics" is addressed later in this chapter, what this section says holds for almost any of its meanings.

The case for reducing the number of different topics taught in science, mathematics, and technology is straightforward. A basic message from research on how children learn science is that (1) many science concepts are inconsistent with children's beliefs about how the natural world works, and (2) for children to understand science concepts often requires that they wrestle with how those concepts are more satisfactory than their own current beliefs. Learning science effectively, therefore, requires direct involvement with phenomena and much discussion of how to interpret observations. Moreover, it requires encountering the intended concepts in a variety of contexts and successively more adequate formulations—activities that obviously take time.

Thinking about Major Topics

Of course there is a trade-off to be made. The argument here is to give up some "coverage" to enable students to gain an understanding of key ideas. According to TIMSS researchers, the state curriculum guides they sampled in 1993 for their study "included so many topics that we cannot find a single, or even a few, major topics at any grade that are the focus of these curricular intentions. These official documents, individually or as a composite, are unfocused. They express policies, goals, and intended content coverage in mathematics and the sciences with little emphasis on particular, strategic topics." (Schmidt, McKnight, & Raizen, 1997). It is true that teachers already regularly eliminate topics from the overload in their textbooks—sometimes by not getting to the final chapters, sometimes by skipping chapters that are too difficult for many of their students (or are troublesome for some community or personal reason). Clearly there are some undeniable limits to how much students can be expected to study even superficially.

"One thing I'll say for us, Meyer—we never stooped to popularizing science."

Deciding what topics to keep and what to give up was the task undertaken by Project 2061 in a three-year study involving hundreds of the nation's leading scientists and educators. Their work resulted in *Science for All Americans*, a statement of the knowledge and skills students should have by the time they graduate from high school. *Science for All Americans* was followed by a four-year study involving an even larger number of scientists and educators that led to *Benchmarks for Science Literacy*, a statement of what students most need to learn as they progress through school. The bar graphs in the box on the following page show the number of ideas included in *Benchmarks* that are also found in the traditional science curriculum that almost everyone is exposed to. Whatever minor uncertainties there may be in the count, it is evident that well over half of the traditional curriculum content was omitted from the recommended core.

In determining what topics to exclude, Project 2061 developed two basic criteria for evaluating candidate topics. A topic was not included in either *Science for All Americans* or *Benchmarks* if there were no compelling argument that it would be essential for science literacy, or if its importance were judged to be out of proportion to the amount of time and effort that would be needed for all students to learn a coherent set of concepts about it.

For example, Project 2061 took a close look at the topic of electrical circuits. This happens to have been the subject of considerable research on students' learning difficulties, in terms of both the necessary input of learning effort and the likely output of fruitful knowledge.

On the input side, how learnable are circuit ideas? Some researchers have spent their careers trying to understand why students—from elementary school to college—have so much difficulty in understanding not just the differences in behavior of series and parallel circuits, but even the very notion of what a circuit is. Even when researchers have thought they understood the nature of students' difficulties and misconceptions, they still have had trouble figuring out how to overcome them. So, at best, a great deal of extra classroom time would have to be spent on getting students to understand electrical circuits.

On the outcome side, how important is it to science literacy for students to understand electrical circuits? The judgment has to be made on the basis of the importance of that knowledge itself, the prior knowledge required to learn it, and what other knowledge it will lead to or support. By itself, electrical circuitry does not have much to offer science literacy. Practical knowledge of electrical circuits may be required for students who will specialize in physics or engineering, and it would also be of value to

Electric Currents

"Grades 6-8: Electric currents and magnets can exert a force on each other.

Grades 9-12: Magnetic forces are very closely related to electric forces and can be thought of as different aspects of a single electromagnetic force. Moving electric charges produce magnetic forces and moving magnets produce electric forces. The interplay of electric and magnetic forces is the basis for electric motors, generators, and many other modern technologies, including the production of electromagnetic waves."
—*Benchmarks for Science Literacy*, p. 95 and p. 97

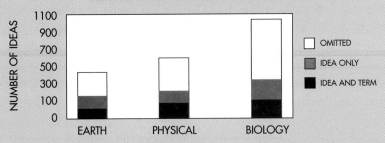

In developing *Benchmarks for Science Literacy*, a count was made of ideas found in a traditional textbook series—physical science, earth science, and biological science—that approximate the science curriculum to which all students are exposed. "Ideas" were defined as boldface subsection headings in the text and/or glossary entries.

Through eliminating, pruning, and trimming (as described in this chapter), fewer than half of these ideas were retained as essential to basic science literacy. The black areas in the bar graphs above indicate the number of instances in which both the idea and the technical term for it were included in *Benchmarks*. The gray areas indicate the number of instances in which *Benchmarks* included the idea, but not associated technical terms judged to be unhelpful to understanding or required for science literacy. The white areas indicate the number of ideas that were not included explicitly in *Benchmarks*.

The reduction in biology is particularly striking. In some areas, *Benchmarks* has effected a wholesale reduction of myriad details in favor of understanding general principles. The most dramatic example deals with the characteristics of different phyla, represented in three grade 6-8 benchmarks:

- One of the most general distinctions among organisms is between plants, which use sunlight to make their own food, and animals, which consume energy-rich foods. Some kinds of organisms, many of them microscopic, cannot be neatly classified as either plants or animals.
- Animals and plants have a great variety of body plans and internal structures that contribute to their being able to make or find food and reproduce.
- Similarities among organisms are found in internal anatomical features, which can be used to infer the degree of relatedness among organisms. In classifying organisms, biologists consider details of internal and external structures to be more important than behavior or general appearance.

Some textbooks devote hundreds of pages to conveying these concepts, but in no way do the benchmarks themselves justify keeping all of this detail (other than as reference material for students studying chosen aspects of differences).

do-it-yourselfers to understand what is happening in, say, a three-way switch arrangement, but even they would be well advised to follow standard wiring diagrams rather than figure it out on their own.

On the other hand, the idea of an electric current plays an important role in science literacy because of its relationship to magnetic fields in electric motors, power generators, Earth's magnetic field, and more. For those links, however, less need be known about currents than is necessary for making sense of series and parallel circuits. The marginal note on page 213 presents the benchmarks for grades 6-8 and 9-12 that are relevant to an understanding of electric currents.

Project 2061 concluded, therefore, that series and parallel electrical circuits as a subject was best left out of the goals for the core science curriculum on the grounds that it would require a high instructional cost and provide a low payoff. Paradoxically, one of the most popular instructional units among elementary- and middle-school science educators is the hands-on science activity "batteries and bulbs," in which students investigate series and parallel circuits. It may be that this engaging activity can be adequately justified by its contribution to understanding scientific reasoning—hypotheses, evidence, modeling, observation, and so on—even if students are not likely to retain knowledge about series and parallel electrical circuits. And of course any student with an interest in electrical or electronics technology ought to have some opportunity outside of the common core to study circuits. In any case, the point here is not to single out conclusions about the topic of electrical circuits for special attention but to illustrate the kind of analysis that is needed in deciding which topics ought to be included and which left out.

A Process for Cutting Topics

There are few substantive differences between what is included in the content recommendations of AAAS (Project 2061's *Benchmarks*) and those of the National Research Council (*National Science Education Standards*) for what all students should learn. Yet neither organization explicitly states which topics can prudently be eliminated from the basic science-literacy core. Rather, they recommend what knowledge and skills are to be learned, leaving it to teachers and curriculum developers to decide which topics to have students study to achieve those learning goals. The following box explores this important distinction further. The process of reaching a consensus on which topics to eliminate is itself an effective way for teachers to clarify those distinctions in their own minds.

Understanding electric circuits is problematic for most students, even in college. In the videotape *Minds of Our Own* (Harvard/Smithsonian), researchers feature recent MIT graduates who are unable to light a bulb with a battery and wire. One student's drawing below illustrates a typical misunderstanding.

There are similar findings for Harvard University and MIT graduates unable to explain photosynthesis, seasons, or molecules.

Designs on Disk can help teachers select topics to be dropped from the core curriculum in science, mathematics, and technology by providing appropriate databases and record-keeping forms.

"TOPICS"

The word "topic" carries several meanings in teaching and distinguishing among them is important for following *Designs for Science Literacy* in general and this chapter in particular. Its ambiguity carries risk for misunderstanding curriculum design.

"Topic" as a category of learning goals. One meaning is a heading in an outline of goals or instructional materials—for example, "Cells." There are a great many different facts, ideas, and principles that could be taught or tested under the topic heading "Cells," and the heading by itself gives little or no clue as to what will be included or what is most important. Yet, in the context of a well-established curriculum, topic headings may imply a particular collection of ideas traditionally included under those headings. The heading "Cells," for example, would typically include the names of nucleotides A, C, G, and T, transfer RNA, and endoplasmic reticulum—none of which is in *Benchmarks for Science Literacy*. At one time, Project 2061 considered deliberately avoiding such familiar headings, in the worry that people would read into them the full list of traditional details, in addition to the specific important ideas that were intended. (That is still a worry.)

Obviously, topic headings are not in themselves pernicious. But unless learning goals get more specific than topic headings, they can seriously undermine less-is-better reform, by allowing everything in the current curriculum to be stuffed back in under one heading or another. For that reason, topic headings, though obviously necessary, are viewed with suspicion by Project 2061.

"Topic" as a context for learning. Another meaning of "topic" is something that students study. When students are asked what they are currently studying in school, their answer is likely to be the topic in this sense. Examples include "lakes," "earthquakes," "environmental pollution," "paper," or "bridges." They may be studying one of these topics, not necessarily because it is important to learn about in itself, but because it provides opportunities to learn and use some ideas and skills that are important. So, for example, in studying "paper making" (to which there are no direct references in *Benchmarks*), students may learn about measurement, experimental design, materials and manufacturing, the side effects of technology, communication, and other learning goals for which *Benchmarks does* indeed specify particular learning goals.

Topics as headings are more likely to be discipline divisions (chemical equations, heredity, or multiplication). Topics as learning contexts are more likely to be phenomena (lakes or earthquakes), events (the Olympics or exploring space), or societal issues (health care or environmental pollution). But some topics can be both – "Cells," "Nuclear Power," and "The Solar System," for example, come with their own benchmarks and also provide opportunities to learn.

There is an important distinction between "topics" and "learning goals."

A "topic" can mean a set of learning goals (such as understanding plant classification) or it can mean a teaching unit (such as "Rainforests"). And of course it often can be both, when a familiar teaching unit has some obviously associated goals—such as a unit named "Electromagnets," which implies both a familiar set of ideas to be learned and relevant activities with batteries, wires, and nails. So "dropping a topic" may mean giving up on a set of expectations for what students will learn, or forgoing a set of customary teaching activities, or both. But whatever the overlap between goals and activities for a particular "topic," the recommendation in this chapter is the same: to hold it up against benchmarks and consider dropping it if its true cost is too high compared to its learning benefits.

To begin eliminating whole topics from the core, faculty teams in science and mathematics for each grade or grade band could be challenged to drop one major science topic and one major mathematics topic from the curriculum in the next semester. The teams should then go through the following step-by-step process:

1. Discuss the distinction between (1) "topics" as categories of goals for what students will end up knowing or being able to do, and (2) "topics" as contexts for learning, rather than content to be learned. (They could consider just not using the word "topic" for one or the other of those meanings, but the word is so firmly embedded in education discourse that it is best to wrestle with it for a while.)

2. Begin with a list of topics in the current curriculum (often textbooks can serve as proxies for the curriculum) and indicate whether there are learning goals in *Benchmarks* that match. The various teams in the district can share and compare lists.

3. As the number of items on the lists grows, begin to make a master list of topics to be considered for elimination from the core. The criteria for making such judgments would, of course, have to take into account any pertinent district or state requirements.

4. From the list of candidate topics for elimination, each team member selects one topic to drop and identifies a core topic in which to invest the additional time made available.

5. Each team member evaluates the effect of dropping one topic and spending more time on another. It may be useful to consult with teachers in later grades about their expectations for what their students should already have learned. (Since those teachers too are struggling with the importance of topics, their advice is desirable but not definitive.) In each instance, after full discussion, the faculty team decides whether to recommend that other teachers in the district also drop the topic in question from the basic science and mathematics core.

The evident overcrowding of the curriculum may be sufficient motivation for many teachers to undertake the kind of topic-reduction process described here. More motivation can be generated by viewing and discussing videotapes in the Annenberg/CPB Multimedia Collection that demonstrate how easy it is to overestimate what students learn.

Ideally, an evaluation of trade-offs associated with cutting topics would also consider what students had learned in the time saved. Good use of freed-up time is far from a trivial task in itself and is brought up again at the end of this chapter.

From the chapter "How to Buy Plants" in The Garden Primer *(Workman Publishing, 1988).*

The five-step process is then repeated, involving more teachers and topics as confidence in the process grows and as the benefit of spending more time on fewer topics becomes apparent—and also as it becomes clear that there are no terrible side effects associated with the process.

This gradual approach to topic elimination is manageable and not terribly risky, since action is based on careful group analysis and is limited in scope. It does not threaten to strip teachers summarily of their favorite topics or to mandate wholesale changes in curriculum content. But it is intended to begin a systematic process of thought and action that stresses basing topic decisions on intended learning outcomes, takes benchmarks and national standards seriously, and tries out ideas on a small scale before making recommendations for districtwide implementation.

It is undoubtedly painful to eliminate familiar, even beloved, parts of instruction. As one teacher said, "I have to teach gas laws. I have always taught gas laws. I like to teach gas laws." Curriculum conservationists often ask, "How can you leave out important topic X?" But, since resources are limited, two other questions also need to be asked: "How long does learning X take?" and "What topics would you like to leave out to make room for X?"

Some educators are against taking any topics at all out of the current curriculum, and indeed, would like to add more, which often requires pushing topics into ever lower grade levels. Project 2061 would hope that the educators make sure that students learn at least the ideas in *Benchmarks.* If they do, the project is happy to have them learn any additional number.

The box opposite lists topics that appear in traditional textbooks yet do not contribute toward students' achieving the learning goals specified in *Benchmarks* or *National Science Education Standards* (*NSES*). Although these topics conceivably could be treated in a way that would serve related benchmarks, they were not treated that way in the textbook they appeared in. There remains the possibility that any one of these topics, as in the instance of electrical circuits, could be studied in a way that would carry benefits for achieving "process" goals—say, those in the *Benchmarks* chapters THE NATURE OF SCIENCE or HABITS OF MIND.

Many of the listed topics could also constitute advanced work for students who understand the basic ideas in the core. But claims of promoting higher-order thinking cannot be used as a justification for any topic whatever without considering the efficiency of the experience. One should ask, for example, how much of the time devoted to batteries and bulbs promotes learning scientific reasoning, how much goes into the

TRADITIONAL TOPICS TO CONSIDER
EXCLUDING FROM LITERACY CORE

Here are some topic headings taken from typical textbooks under which few
(if any) relevant benchmarks could be identified:

from a typical **Physical Science** textbook

Gas Laws	Flight	Electric Circuits
Periodic Table	Work & Power	Optics
Properties of Solutions	Simple Machines	Nuclear Reactors
Acids & Bases	Calorimetry	Mining
Nuclear Chemistry	Heating Systems	Petroleum Processing
Buoyancy	Refrigeration Systems	Electronics
	Engines	Computer Hardware

from a typical **Earth Science** textbook

Solar Features	Lunar Features	Rivers
Stellar Evolution	Atmospheric Layers	Geological Eras

from a typical **Biology** textbook

Branches of Biology
Classification System

from typical **Algebra** and **Geometry** textbooks

Rational Expressions	Fractional Equations	Axiomatic Systems
Conic Sections	Quadratic Inequalities	Locus
Matrix Operations	Systems of Inequalities	Synthetic Methods
Polynomials		Right Triangle Trigonometry
Factoring		
Radical Expressions		Sets and Truth Tables

Another possible response to this information is to claim that benchmarks and standards are themselves lacking—for example, that simple machines should be part of science literacy. Local educators should always have the option of setting different priorities (though they would be well advised to consider what other ideas should be neglected to make room for simple machines). The main point here is not that the current version of *Benchmarks* is inviolable, but that the argument for including questionable topics has to be linked to specific learning goals, not rest on the mere familiarity of topics.

fundamental notion of a circuit, and how much goes only into a fruitless struggle with series and parallel principles—and whether processes of science could be learned just as well by studying a more important topic.

It should not be inferred that the topics in the list on the previous page are absent altogether from *Benchmarks*. For example, there is nothing explicit in *Benchmarks* about the periodic table as such, but there is the notion of periodicity in elemental properties—that is, that there are families of elements with similar properties and that similar sequences of properties appear when elements are arranged in order of their atomic mass. Similarly, there is nothing about gas laws in their symbolic-quantitative form, but the benchmarks on temperature and molecular motion would likely require experience with compressibility of gases and their increase in pressure when heated.

It is evident from the list that, at least in terms of the traditional way of organizing textbooks, low-priority *major topics* are distinctly easier to identify in physical science than in the more interconnected life science. In the next section, it is evident that low-priority *subtopics* are easier to identify in life science.

PRUNING SUBTOPICS FROM MAJOR TOPICS

Similar arguments can be made for a less radical adjustment of traditional curriculum content that will leave time for higher-priority learning goals. Part of the curriculum problem is that, in addition to treating too many major topics, the curriculum treats many subtopics within them with excessive detail (relative to the topic's importance for literacy). In addition to eliminating whole topics, therefore, progress can be made by cutting back on the extent and complexity of the treatment of at least some topics. Whereas dropping whole topics can lead to the elimination of whole chapters or units, pruning may correspond loosely to cutting out paragraphs at the subtopic level. The purpose of such pruning is to focus on what is really important to know about a topic rather than on how to make it easier to learn.

The following four tables suggest subtopics (for physical, earth, and biological science and for algebra and geometry) that could be considered for pruning from the lists found in a typical set of textbooks intended for all students in grades 8–10. The textbooks themselves (subject to weight limits) could contain all these ideas for students going beyond basic literacy—and as a reference for all. But a full understanding of the most important ideas first will facilitate learning these extras.

It is important to remember that Project 2061 does not claim these topics to be unimportant, only less important than those with higher priority for basic literacy in the limited time available in school.

SUBTOPICS TO CONSIDER FOR PRUNING

Subtopics in a typical **Physical Science** textbook under which
few (if any) relevant benchmarks could be identified

Atomic Structure
Thomson's model
Rutherford model
Bohr model
mass number
shell filling
quarks

Chemical Reactions
network solids
metallic bonds
oxidation number
single replacement
double replacement

Organic Chemistry
structural formulas
isomers
saturated hydrocarbons
alkanes, alkenes,
   alkynes, cycloalkanes
aromatic hydrocarbons
substituted hydrocarbons
alcohols & hydroxyl
   group
organic acids & carboxyl
   group
esters & esterification

halogen derivatives
lipids

Force & Motion
conservation of momen-
   tum
sliding vs. rolling fric-
   tion
free fall
inclined planes

Electrostatics
electrostatic induction
electric discharges
grounding

Electric Circuits
volts, amperes, ohms
Ohm's law
electrochemical cells
electrodes
electrolyte
thermocouple
alternating & direct
   current
series & parallel circuits
fuses & circuit breakers

Magnetism
magnetic lines of force
magnetic induction
temporary magnets
magnetic variation
magnetic domain
transformers

Sound
intensity, quality, timbre
decibels
fundamental &
   overtones
resonance
reverberations

Light
polarized light
photoelectric effect
index of refraction
primary & complemen-
   tary colors
primary & complemen-
   tary pigments
incandescent &
   fluorescent
phosphors
neon light

SUBTOPICS TO CONSIDER FOR PRUNING

Subtopics in a typical **Earth Science** textbook under which
few (if any) relevant benchmarks could be identified

Ocean Features
variable salinity
surface zone & deep
  zone
continental margin &
  shelf
turbidity
fringing reefs
atolls & barrier reefs
intertidal zone
neritic, bathyl, abyssal
tsunamis

Hydrosphere Features
valley & continental
  glaciers
zone of saturation
zone of aeration

Lithosphere Features
coastal & interior plains
primary & secondary
  waves
composition of mantle
Mohorovic layer
asthenosphere
hanging & foot walls
normal & reverse faults
thrust & lateral faults
fault-clock mountains
anticlines & synclines

isostasy
surface waves
volcanic dust, ash,
  & bombs
shield & composite
  volcanoes
Pangaea
transform faults
divergent boundary
convergent boundary
strike-slip

Rock & Soil
streak
cleavage
extrusive & intrusive
  rocks
chemical rocks
stable rock
plant acids
carbonation
pore spaces
soil profile
subsoil
loess

Climate
sea & land breezes
doldrums
trade winds
prevailing westerlies

polar easterlies
anemometer
microclimates
glacial & interglacial
  periods

Fossils
molds, casts, & imprints
trace & index fossils
unconformity
intrusions & extrusions
varves

Fuels & Environment
peat
types of coal
petrochemicals
photovoltaic cells
geothermal
biomass & gasohol
contour plowing &
  terracing
strip cropping
desertification
desalination
temperature inversion
acid rain
catalytic converters
point and nonpoint
  sources

SUBTOPICS TO CONSIDER FOR PRUNING

Subtopics in a typical **Biology** textbook under which few (if any) relevant benchmarks could be identified

General
steps for scientific method
branches of biology
specific microscopes
limits of resolution
lab techniques of biologists
chemical vs. physical
  change

Organic Chemistry
dehydration synthesis
hydrolysis
lipids & saturation
phospholipids & cholesterol
peptides & peptide bonds
nucleic acids & nucleotides
RNA = ribonucleic acid

Cell Structure
nucleolus
vacuoles
plastids
osmosis
facilitated diffusion
active transport
endoplasmic reticulum
Golgi apparatus
lysosome

Cell Energy
photosystems
electron transport

ATP formation
dark reactions
Calvin cycle
glycolysis
Krebs cycle
anaerobic energy
  production
lactic acid fermentation
alcoholic fermentation

DNA & Protein Synthesis
structure of DNA
base pairing
replication of DNA
RNA and its structure
transcription & translation
A, C, G, T
double helix

Cell Growth & Division
rates of cell growth
controls of cell growth
phases of mitosis
cytokinesis
chromatin
chromosome structure
centriole & spindle

Genetics
self- & cross-pollination
dominant & recessive
segregation

F1 cross
Punnett square
homozygous &
  heterozygous
two-factor cross F1 & F2
phases of meiosis
crossing over

Genes & Chromosomes
chromosomes
linkage groups
sex linkage
X, Y chromosomes
sex determination
sex-linked genes
chromosome mutation
  types
point mutations
frameshift mutation
dominance & codominance
polygenic inheritance
operon & operator
promoter, inducer, repressor
eukaryote gene expression
exons & introns

Human Genetics
human blood groups
Huntington disease
sickle cell anemia
sex determination
sex-linked genetic disorders

sex-influenced traits
chromosomal abnormality
genome

Ecology
climax community
nitrogen fixation & cycle
denitrification

Populations
"exponential" growth curve
logistic growth curve
carrying capacity
density-dependent limiting
density-independent limiting

Evolution
eras, epochs, & periods
gaps in fossil record
quality of fossil preservation
relative & absolute dating
reproductive isolation
gradualism
mass extinction
adaptive radiation
divergent & convergent
early atmosphere
microfossils
anaerobes

SUBTOPICS TO CONSIDER FOR PRUNING

Subtopics in typical **Algebra** and **Geometry** textbooks under which
few (if any) relevant benchmarks could be identified

Rational Expressions
simplifying rational expressions
rationalizing the denominator
operations on rational expressions

Factoring
factoring polynomials
solving quadratics by factoring

Matrices
operations on matrices

Polynomials
multiplication of polynomials
division of polynomials
FOIL method

Radicals
simplifying radical expressions
operations with radicals
solving radical equations

Logarithms
solving logarithmic equations
converting bases

Axiomatic Systems & Proof
incidence & betweenness theorems
two-column proofs
synthetic methods

Circles
theorems about ratios of segments,
tangents, and chords
theorems about angles formed by
tangents and chords

Vectors
operations

Sets
Venn diagrams
union & intersection operations and
properties

Angles and Polygons
alternate exterior angles
oblique polyhedra
polyhedral angles
reflex angle
trapezeum
Heron's formula

Thinking about Subtopics

With regard to electric circuits, for example, one could give up on the time-consuming distinction between series and parallel, on the quantitative relation $I=V/R$, and even on the distinction between voltage and current, and focus instead on the centrally important principle that electric circuits require a complete conducting path. (The complete circuit idea is difficult enough in itself, as the ample research on students' electrical misconceptions shows.) Similarly, in teaching about DNA and protein synthesis, it should be enough to concentrate on helping students to understand that cells construct proteins according to instructions coded on DNA molecules without teaching them about introns, exons, A, G, C, T nucleotide codons, and messenger, transfer, and ribosomal RNA—details that are found in nearly all introductory high-school biology textbooks and many middle-school ones. Although a few students are fascinated by and eager to learn about the complexity of the actual mechanism, many more students are intimidated by it; they barely remember the details long enough for a test and, even worse, never get the general idea at all.

The optimistic notion that students may forget the details but remember the basic idea is rarely supported by research findings. Yet some details are surely necessary to make the basic ideas intelligible and plausible in the first place. How many and which details, and how they are best tied to the basic ideas, are issues that are waiting to be demonstrated by more focused research on learning. It seems likely that the optimum solution will include frequent reminders of how the details relate to the big ideas.

Science for All Americans and *Benchmarks for Science Literacy* are as concerned with the level of understanding of topics in science, mathematics, and technology as they are with which particular topics are essential. The Project 2061 recommendations have been painstakingly worded to signal the level of understanding that is sufficient for purposes of general science literacy, and they agree extensively with the *National Science Education Standards* on what those expectations are. Of course, many students will go further and deeper into at least some topics.

The general similarity in pruning topics implied in *Benchmarks* and in *NSES* sends a strong signal from the scientific community that teachers would do well to focus on student understanding of key ideas. The reports do not argue against the introduction of detail in teaching, nor argue in favor of having students memorize only vague generalizations. Just the opposite. Teaching should present key topics with enough concrete detail and hands-on involvement to make them interesting and memorable, but not with so much that the main ideas are obscured and that students believe that memorizing a collection of details or carrying out a collection of steps constitutes understanding those ideas.

Resources for Science Literacy: Professional Development compares *Benchmarks for Science Literacy* and *National Science Education Standards* thoroughly. The summary of the comparison lists differences in what they recommend—and therefore differences in what they imply can be pruned out.

An issue related to what traditional topics to prune is what traditional experiences to prune. Some experiences (say, planting beans, or measuring volume by displacement of water) have as fixed a role in the traditional curriculum as some topics. Again, the Project 2061 position is that time in the curriculum has to be justified by what students learn. Eventually, well-designed curriculum blocks will solve many of the problems of efficacy and efficiency. But in the meantime, all components of instruction can be questioned and experimented with.

And how much time should be allowed for students to invent or discover ideas for themselves? Although there is little research to suggest "discovery" learning is actually more effective, most educators wish that more time were available for hands-on experimentation and student inquiry. To the extent that a set of specific learning goals is believed to be important, however, the time cost of discovery is a serious limitation. The more time spent on discovering any one idea, the less time there is available for learning all other ideas. Almost all educators agree that there should be a balance between how much is learned and how well it is learned—and that there should be at least enough discovery, inefficient as it may be, for students to learn what it feels like.

A Process for Pruning Subtopics

A modification of the process outlined above for cutting major topics can be used to reduce those that are retained. Grade-related faculty teams should be formed in science and mathematics to find subtopics that can be pruned away and to suggest what specific material to remove. The teams should start by exploring the relationship between specific goals and the study topics intended to target them. Eventually, teams should build a school-district database of topics that have been tightened up in terms of how well they target specific learning goals.

At some point in their deliberations, the teams should begin to formulate a list of topics to be considered as candidates for pruning. Then each team selects a topic from the candidate list for intense examination. Working in small groups and using either *Benchmarks* or *NSES* for guidance, the team members should decide which details can be removed or how the topic can otherwise be simplified without limiting its ability to serve the identified learning goals. Of course the team members have to make sure they do not cut material that is necessary for the understanding of some other idea in the same or later grade ranges. Inviting consultation (but not veto power) from teachers at higher grade levels can be a useful part of the process.

The teachers on the teams may then volunteer to test the slimmed-down topic in

Students often have to wrestle with new ideas that are inconsistent with the ideas they already have; experiences that evoke and resolve that struggle can often be arranged for students, without requiring that they discover the new ideas themselves.

A utility on *Designs on Disk* makes it possible to identify, for any subtopic being considered for pruning, which *Benchmarks* concepts may depend on it.

their classrooms using the time saved to ensure better learning of that topic. They should keep notes on what happens so they can answer such questions as: What effect did pruning have on student interest in the topic? Did the students learn the essential ideas or skills? How much time was actually saved? What changes should be made next time around to improve the quality of the learning? Once the treatment of a topic has been worked out to the satisfaction of the team, it should be written up and entered in a *Designs on Disk* file and made available to all teachers in the district.

The pruning of details from a topic may well be more difficult than the elimination of major topics that can more or less stand alone. The details may be so woven together that the instructional strategy unravels when some threads are pulled out. (In the Project 2061 curriculum-materials evaluation procedure, one of the criteria for describing a material is how easily the core ideas can be distinguished from extended detail—and then how easily the instruction for the two can be separated.) Close attention has to be paid to how well the closely surrounding instruction holds together when some aspect is deleted.

TRIMMING TECHNICAL VOCABULARY

A special case of pruning topics involves cutting back on the teaching of technical terms for their own sake. It is not an easy task. Some teachers say that technical vocabulary has been an integral part of their instruction for so long that they can barely conceive of what topics would be without it. And de-emphasizing vocabulary may not produce immediate cheers from students either, particularly the older ones, since many of them have come to believe that memorizing words is the same thing as understanding the concepts—and they have become very good at it. Students' inclination, reinforced over years of school-ing, to substitute memorization for understanding is all the more reason for teachers to

One of the 25 criteria that Project 2061 uses for judging the quality of instructional materials is labeled Introducing Terms. This is a very short statement of it.

Criterion: Introducing Terms

Does the material introduce technical terms only in conjunction with related experience and only as needed to facilitate thinking and promote effective communication?

Clarification. Terms are important to scientific communication. For students to be able to communicate efficiently about phenomena, some technical terms are helpful. However, concentrating on vocabulary rather than on understanding carries the risk of leading both students and teachers to mistake fluency with terms for understanding. It also mistakenly portrays science as learning "big words" rather than as asking and answering questions about the world. Understanding, rather than vocabulary, should be the main purpose of science teaching and hence, the number of terms used should be limited to those that are essential for communicating about experiences.

Responding to this criterion involves examining whether the material (1) introduces technical vocabulary mainly in conjunction with experiences and (2) limits the use of technical terms to those needed for communication about those experiences.

help students get better at learning content that has greater utility and durability.

Take, for example, the topic "Cells." For some teachers, its importance justifies having students learn the names of the parts of the compound microscope, copy drawings of generalized cells from the book, or learn to spell "endoplasmic reticulum," "mitochondrion," "organelles," and "cytogenesis." This is not to disparage the topic of cells itself, for indeed it is emphasized by both the *NSES* and Project 2061's *Benchmarks.* Nor is it to downplay the need for correct spelling. But as the figure on the facing page shows, there is a vast difference between the language Project 2061 uses to express what students need to know about cells and the language used in traditional textbooks and hence in curricula. Where one popular high-school biology textbook uses 120 cell-related technical terms, *Science for All Americans* uses but 11.

Thinking about Technical Vocabulary

In *Benchmarks,* as in *Science for All Americans,* there is a strong tendency to avoid using a specialized vocabulary. Once students can explain that cells get energy from food and use the energy to put together complex proteins, their knowing such terms as "oxidation," "respiration," "mitochondrion," and "ribosome" can be helpful, but learning the words without the basic notion is empty. (Could the words be learned first and then later be put together meaningfully? Probably so—but all too often the learn-

In the lower grades, *Benchmarks* uses simpler language than *Science for All Americans,* in an attempt to characterize what all students at each level may be expected to understand.

CELL TERMINOLOGY

cytoplasm electron transport chain PROTEIN lysosomes
GENE energy gradient organelles
RDP hydrolysis NUCLEUS CATALYST PGAL
CHROMOSOMES glycolysis substrate chromatin
pyruvic acid peptide bond nuclear envelope
pyruvic acid conversion monomer nuclear pores
aerobic respiration polymer CELL MEMBRANE
anaerobic polysaccharide nucleolus acetyl-CoA
CELL WALL Krebs cycle middle lamella
interphase endoplasmic reticulum prophase
ribosome cytokinesis Golgi bodies metaphase
mitochondria anaphase plastids telophase
vacuoles cell plate homologous chromosomes
microtubules diploid spindle fibers
haploid centrioles somatic cilia mitosis
flagella meiosis eukaryotic polar body
prokaryotic REPLICATION
binary fission chloroplast vegetative propagation
leukoplast regeneration contractile vacuoles
gametes facilitated diffusion SPERM active transport
ovum carrier molecule zygote hypertonic solution
centromere hypotonic solution synapsis
solute tetrad solution DNA
RNA concentration gradient nucleotide
endocytosis adenine turgor guanine
exocytosis thymine phagocytosis cytosine
pincocytosis purines ADP pyramidines ATP
deoxyribose ENZYME ribonucleic acid
phosphate group ribose cellular uracil
PHOTOSYNTHESIS messenger RNA xanthophylls
transfer RNA carotenes ribosomal RNA grana
transcription stoma codon

Here is a list of technical terms found in two chapters on cells in a typical high school biology textbook. Terms that appear in color are those that also appear in *Science for All Americans* and *Benchmarks for Science Literacy*.

The intention of the criterion for introducing terms is evident in these verbatim passages from middle-school textbooks. In the passage from textbook A, the term "photosynthesis" appears abruptly, without preparation. (There are accompanying activities on food webs but not photosynthesis.) In the passage from textbook B, the term appears only after relevant phenomena have been studied, and the basic process has been described in plain English.

TEXTBOOK A
ENERGY FLOW

Every community of living things, whether in a forest or in a city, needs energy to support life. You know that an automobile gets its energy from burning a fuel, usually gasoline. Well, a community of living things also gets its energy from "burning" a fuel, but it gets energy in a different way and from a different fuel—food.

Think about how energy "moves" through a forest community. Some animals get their energy (or food) by eating other animals. Some animals eat plants. But what do plants eat? What is their source of energy?

ENERGY → PLANTS → ANIMALS → ANIMALS

A few plants, such as the pitcher plant and the Venus' flytrap, eat animals (insects). But this is not the ususal niche that plants occupy. So how do plants usually get their energy? During **photosynthesis,** green plants change carbon dioxide (a gas they get from the air) and water (from the soil) into food. This food supplies energy not only for the plants, but also for the animals that eat the plants. The process of photosynthesis also *requires* energy. Green plants use energy from sunlight to manufacture their food, as you can see from the following energy diagram.

SUN'S ENERGY → PLANTS → ANIMALS → ANIMALS

TEXTBOOK B
PUT IT ALL TOGETHER!

So now what do you think about how plants get food? If you decided, as a result of doing these actvities, that plants use light, water, and carbon dioxide to make starch, you are right. Plants actually *make* their own food in their leaves. To do this, they need energy from the sun plus two raw materials: water and carbon dioxide. The process by which plants make food is called **photosynthesis**. Like many scientific terms, this one is a combination of two Greek words: *photo*, which means "light," and *synthesis*, which means "putting together." Given what you learned in doing the activities, why do you think the process is called *photosynthesis*?

Ultimately, photosynthesis is responsible for feeding practically all of the organisms on Earth! Plants use photosynthesis to make their own food. Animals feed directly on plants, and other animals feed on those animals. At your next meal, think about the fact that none of the food on your plate could exist without photosynthesis.

ing stops well short of understanding.) And in the end, the idea is understandable without those terms to a degree sufficient for general science literacy.

Technical language is helpful when the same idea needs to be referred to again and again. If there is a legitimate reason to refer to where food is oxidized in a cell, it obviously doesn't make sense to endlessly repeat "that special part of the cell where food is oxidized." Although Project 2061 recommends minimizing unnecessary technical terms, it is committed to expanding students' useful scientific vocabulary. The correct use of technical vocabulary by students is to be applauded once they understand the meanings. But as a rule of thumb, understanding should come first, definition after. (See the facing page for contrasting examples of how two textbooks introduce technical terms.)

But why not encourage the use of poorly understood terms, just to get students accustomed to the use of technical language and develop in them a sense of having learned something special? In fact, isn't that how we ordinarily build vocabulary—first only dimly understanding new words and then refining their meaning with time? Perhaps so, but it is clear that in school many negative consequences come from the premature use of technical vocabulary—not the least of which is the persistent impression that "science" means having mysterious names for everything. "Evaporation" is an excellent example of a term that children often are taught to say long before they have any idea of what a vapor is, or even know that air is a substance. Although a child's use of "evaporates" may signify no more than a fancy word for "disappears," some listeners are prone to interpret it as evidence of understanding kinetic-molecular theory.

The pressure to cover the curriculum and test students efficiently makes many teachers, administrators, test makers, and parents—not to mention materials developers and textbook publishers—too willing to interpret students' glib use of technical terms as evidence of understanding. As a result, learning is short-circuited and students are not able to increase their sophistication gradually. Teachers and test makers should require students to explain what they mean, not just come up with the right word. (For example, the question should not be "What do we call the part of a cell that does x?" but rather "What processes in a cell have special parts to perform them?" Only then, perhaps for *extra credit*, "What are those parts called?")

Another reason for de-emphasizing technical vocabulary is to concentrate attention on what ideas are required for literacy. If a goal stated that students "should become familiar with kinetic-molecular theory," many readers would be satisfied. But what has really been recommended? Knowledge of the ideal gas laws? The vectorial argument for equipartition of energy? If instead it is said that students should know

SOME EXAMPLES OF HOW *BENCHMARKS* USES TECHNICAL VOCABULARY

Both the term and the concept are included in *Benchmarks*:

feedback
"The feedback of output from some parts of a system to input of other parts can be used to encourage what is going on in a system, discourage it, or reduce its discrepancy from some desired value." — *Benchmarks* section SYSTEMS

ecosystem
"Like many complex systems, ecosystems tend to have cyclic fluctuations around a state of rough equilibrium." — *Benchmarks* section INTERDEPENDENCE OF LIFE

electromagnetic
"The interplay of electric and magnetic forces is the basis for electric motors, generators and…the production of electromagnetic waves." — *Benchmarks* section FORCES OF NATURE

The concept, but not the term, is included in *Benchmarks*:

photosynthesis
"Plants use the energy from light to make sugars from carbon dioxide and water." — *Benchmarks* section FLOW OF MATTER AND ENERGY

entropy
"In any interactions of large numbers of atoms and molecules, the statistical odds are that they will end up with less order than they began with — that is, with the energy spread out more evenly.
"Transformations of energy usually produce some energy in the form of heat, which spreads around by radiation or conduction into cooler places. Although just as much total energy remains, its being spread out more evenly means less can be done with it." — *Benchmarks* section ENERGY TRANSFORMATIONS

quantum
"When energy of an isolated atom or molecules changes, it does so in a definite jump from one value to another, with no possible values in between. The change in energy occurs when radiation is absorbed or emitted, so the radiation also has distinct energy values." — *Benchmarks* section ENERGY TRANSFORMATIONS

Neither the concept nor the term is included in *Benchmarks*:

osmosis
(Yes, this is important in transpiration and cell maintenance. The term is commonly used, but as a synonym for diffusion.)

black hole
(Black holes are very bright. Higher priorities, necessary to making sense of it, are why things orbit other things and why ordinary stars shine.)

laser
(Everyone knows it produces a very intense light. Explaining it is something else.)

that "everything is made of invisibly tiny pieces that are continually moving and banging into one another, and that this view explains many diverse natural phenomena," far more is contributed to the readers' understanding of what students need to know. In all of this, therefore, the purpose is to take care that fancy language does not supplant or get in the way of plain language used to state learning goals clearly. (Goal statements also imply, of course, what language is reasonable to expect of all students.)

A Process for Trimming Technical Vocabulary

Again, faculty teams can be set up to look for topics that seem to be overburdened with technical language and recommend which terms can reasonably be avoided. As suggested in the box on the opposite page, a list of topics can be considered to include three categories: (1) the concept and the technical term for the concept are both recommended for basic science literacy; (2) the concept is recommended, but not the technical term; (3) neither the concept nor technical term is recommended. As the faculty teams study the terminology associated with a topic, they can question the judgment expressed in the list and modify it if they wish, but only after discussion and then only if persuasive arguments for including the technical terms are made. The burden is on having to show why technical words that go beyond what is needed for science literacy should be included, not on having to argue for their exclusion.

As before, teachers should try teaching and assessing the selected topics without the terms that have been tentatively placed on the list of nonessential technical terms. That involves making sure that students know which words are not required for them to earn satisfactory scores on any tests. Instruction should focus instead on what the phenomena are, explaining them in familiar terms, what their importance is, and where else they will show up, but only incidentally on the technical term to use—when talking to other people who know the technical term. Each participating teacher should write an evaluation of the experience for the investigative team.

Sharing this information, the teams can collaborate in making districtwide recommendations, and the process can continue the progressive reduction of emphasis on memorizing technical terms.

REDUCING WASTEFUL REPETITION

Overloading the curriculum with topics, overloading topics with detail, and having students learn words and terms they don't need are not the only ways to waste

Designs on Disk contains three databases of topics and technical terms found in representative science and mathematics textbooks for different grades. Based on the recommendations in *Science for All Americans, Benchmarks for Science Literacy,* and the *National Science Education Standards*, the databases suggest which topics and terms all students need to learn and which are not essential.

Weighing the helpfulness of technical terms is one step in Project 2061's procedure for analyzing curriculum materials, which asks whether a material:
- Provides a sense of purpose
- Takes account of student ideas
- Engages students with phenomena
- Develops and uses scientific ideas
- Promotes student thinking
- Assesses progress along the way
- Promotes other benefits

instructional time. Another waste is the unnecessary repetition of topics—the same ideas in the same contexts, often with the same activities and the same questions. But deciding what is necessary and what is not is not always an easy matter. The common student complaint that the same topics appear in successive grades, often in the same way, is matched by the common teacher complaint that the students did not learn what they were supposed to before, and so previous topics have to be "reviewed" or, to be frank about it, taught all over again. This situation leads to frustration on the part of both teachers and students and to the loss of opportunities to take up other topics or the same topic in a new and more advanced context.

Considering Redundancy

On the other hand, deliberately revisiting the same concepts can be a valuable instructional strategy. Students' understanding of new ideas does not typically occur in single instructional bursts, but grows gradually over time, often with setbacks and the appearance of new misunderstandings. Many important ideas do have to be revisited in different contexts at different levels of sophistication before they are grasped well by students. But the curriculum is rarely thought out over spans of many grades, and so there is little opportunity to rationalize it by deliberately tailoring repetition.

The understanding achieved by students does not always proceed topic by topic but comes through their assembling ideas from quite different areas. Making sense of the fossil evidence for evolution, for example, requires putting together knowledge of soil erosion and sedimentation, the radioactive transformation of elements, rates of radioactive decay, variation within species, changing proportions, geological change in climate, DNA control of development, and so on. Identifying, coordinating, and sequencing all those components requires an investment of time and cooperation across grade levels that is available only through districtwide support and planning.

A Process for Reducing Unnecessary Repetition

One approach to the problem of wasteful repetition is to use the Project 2061 strand maps found in *Atlas of Science Literacy*. The maps represent the kind of thinking required for planning the fruitful repetition of topics over 13 years of school. The maps illustrate which ideas within and across conceptual strands need to precede others and converge to yield student growth of understanding over the years.

Cross-grade teacher teams should conduct an informal survey to see which topics appear to be frequently repeated in the K-12 curriculum. Choosing from the resulting

A few maps also appear in *Benchmarks on Disk* and on the Project 2061 Web site at www.project2061.org.

list of topics likely to have excess redundancy, the teams should analyze one for which a growth-of-understanding map exists. By comparing entries on the map to the profile of the topic given in the district curriculum, the team can locate differences. By studying these differences, the team can think through just what component ideas may be learned and when, with what level of understanding, and with what means to demonstrate that understanding. Note that this recommendation is for a collaborative investigation of what it takes for students to learn, not for high-school teachers to instruct middle-school teachers, nor for middle-school teacher to instruct elementary-school teachers, in what they should accomplish with students.

THE CHALLENGE

Before wholesale easing of the curricular burden can be attempted or accepted, educators will have to believe that reducing the number of topics, pruning ideas within topics, cutting technical vocabulary, and avoiding needless repetition are worth doing and possible. The small-scale team efforts described above are all within reach. The first steps can improve the curriculum and help colleagues become better disposed to and ready for reform on a more ambitious scale. The more such experiences educators can share in conversation, at conferences, and in newsletters and journals, the closer they can get to a critical mass for the systemic and lasting improvement of science education.

The main point of this chapter has been to make time for teaching the most important ideas more successfully. But knowing how to expand the treatment of a smaller set of topics is not a trivial challenge. To some extent, all teachers know places where there is not enough time to do what they know needs to be done. In other places, it is not at all clear what additional instruction should be done, if more time were made available for it. Even so, it is important to avoid the temptation to include new topics or more details on included topics rather than stretching to improve students' understanding of the core ideas.

In the long run, the effective use of time may be the responsibility of block developers, and teachers' curriculum role will be chiefly to choose an appropriate array of curriculum blocks wisely. In the short run, guidance on how to use time more effectively can be found in the criteria for evaluating lessons and materials found in *Resources for Science Literacy: Curriculum Materials Evaluation.*

Gerald Murphy, *Watch*, 1924-25

Walking Lion in Relief, a Babylonian panel from the 6th century B.C.

CHAPTER 8
INCREASING CURRICULUM COHERENCE

In many school districts, the subjects making up a curriculum in any one year rarely have much to do with one another in practice, even if they do in some abstract description of the curriculum. The treatment of a subject in a given year has little to do with its treatment the prior year or with what its treatment will be the following year. Some topics appear year after year at about the same cognitive level and in the same context, whereas some important topics may never show up in any year at all. Perhaps this lack of intellectual and developmental coherence is not surprising since few K-12 curricula have been designed on the basis of a comprehensive, interconnected set of learning goals—not even within subject areas and certainly not across the entire array of subjects.

This chapter begins by looking at the idea of curriculum coherence and then goes on to discuss ways in which developmental and intellectual coherence of existing K-12 curricula can be improved. Throughout, the context is science literacy and the examples are mostly taken from the work of Project 2061, but there is no reason to believe that the ideas and procedures would not apply more generally.

TYPES OF CURRICULUM COHERENCE

In the context of this chapter, a coherent curriculum is one that focuses on the relatedness of particular knowledge and skills needed for science literacy, takes developmental considerations into account in deciding on the grade placement of specific learning goals in science, mathematics, and technology, and provides occasions for exploring thematic connections between science-related subjects and other fields. These three aspects of curriculum coherence—literacy goals, developmental sequence, and thematic connections—are discussed below.

> "Despite the development of national and state standards, and of standards-driven curricula and tests, the problem of superficial textbooks is still with us and appears to be getting worse, at least temporarily. But it should be said at the outset that mere topic reduction would not lead to higher levels of student achievement. The goal is not merely textbooks with fewer topics, or even lengthier treatment of "key" topics, but books with a coherent vision of the disciplines presented as an unfolding story, allowing even children in the early grades to connect the bits and pieces to larger concepts."
>
> –Harriet Tyson, *Overcoming Structural Barriers to Good Textbooks*, 1997

The International Technology Education Association is working on a set of learning goals for technological literacy.

Literacy Goals

It is probably true that in every field there is more we would like students to learn than there is time for them to learn—even in the thirteen years from the start of kindergarten to the completion of high school. Surely this is the case with regard to science, mathematics, and technology. Hence great care must be taken to identify a limited set of essential, mutually supportive science-literacy goals to serve as the basis for making curriculum-content decisions. Reaching agreement on a *coherent set* of fundamental science literacy learning goals that can withstand critical scrutiny turns out, however, to be demanding, expensive, and time consuming. Fortunately, school districts do not have to do this for themselves, since the basic work has been carried out in a decade-long effort led by three distinguished organizations—the American Association for the Advancement of Science, the National Research Council of the National Academy of Sciences, and the National Council of Teachers of Mathematics—and involving dozens of other scientific societies and education associations.

Project 2061 of the American Association for the Advancement of Science began its work in 1985. It believed that K-12 reform in science, mathematics, and technology education leading to science literacy for all graduates would remain aimless and hence stalled unless a compelling vision of what constitutes science literacy could be formulated. The project mobilized natural and social scientists, mathematicians, engineers, and educators in pursuit of such a formulation, and the result was *Science for All Americans*. From the outset, the broad scope of natural and social sciences, mathematics, technology was chosen in the belief that these areas are so closely connected that the goals for any one of them ought to be conceived in the context of the others.

To be candidates for inclusion in *Science for All Americans*, ideas had to be justified in terms of one or more of the five fundamental criteria of significance and utility described in Chapter 1: employment, citizenship, cultural salience, philosophical reflection, and enriched experience. Moreover, to be included, an idea also had to fit into a set of ideas that would be mutually supporting and together maximize understanding and serve as a foundation for further learning. It is important to recognize that the understanding of science, mathematics, and technology expressed in *Science for All Americans* is not merely a *collection* of ideas, but a *network* of related ideas intended to optimize understanding and further learning of how the world works. Recognition of that relatedness is also part of science literacy. The box on the facing page shows an example of mutually supporting ideas in *Science for All Americans*.

A NETWORK OF RELATED IDEAS IN *SCIENCE FOR ALL AMERICANS*

Chapter 5: The Living Environment, Flow of Matter and Energy

"The accumulating layers of energy-rich organic material were gradually turned into coal and oil by the pressure of the overlying earth. ...Our modern civilization depends on immense amounts of energy from such fossil fuels recovered from the earth." (p. 67)

Chapter 4: The Physical Setting, Energy Transformations

"Forms of energy can be described in different ways... gravitational energy lies in the separation of mutually attracting masses...." (p. 50)

Chapter 9: The Mathematical World, Symbolic Relationships

"There are many possible kinds of relationships between one variable and another. A basic set of simple examples includes (1) directly proportional (one quantity always keeps the same proportion to another), (2) inversely proportional (as one quantity increases the other decreases proportionally)...." (p. 133)

Chapter 4: The Physical Setting, The Earth

"Everything in the universe exerts gravitational forces on everything.... Gravity is the force behind the fall of rain, the power of rivers, the pulse of tides; it pulls the matter of planets and stars toward their centers to form spheres, holds planets in orbit, and gathers cosmic gas and dust together to form stars. Gravitational forces are thought of as involving a gravitational field that affects space around any mass. The strength of the field around an object is proportional to its mass and diminishes with distance from its center." (p. 55)

Chapter 1: The Nature of Science, The Scientific World View

"Science also assumes that the universe is, as its name implies, a vast single system in which the basic rules are everywhere the same.... For instance, the same principles of motion and gravitation that explain the motion of falling objects on the surface of the earth also explain the motion of the moon and the planets." (p. 2)

Chapter 10: Historical Perspectives, Uniting the Heavens and Earth

"Using a few key concepts (mass, momentum, acceleration, and force), three laws of motion (inertia, the dependence of acceleration on force and mass, and action and reaction), and the mathematical law of how the force of gravity between all masses depends on distance, Newton was able to give rigorous explanations for motion on the earth and in the heavens." (p. 149)

Goals set by national organizations for mathematics and science can be found on the Project 2061 CD-ROM, *Resources for Science Literacy: Professional Development.*

Science for All Americans was published in 1989, as was *Curriculum and Evaluation Standards for School Mathematics*, a formulation of learning outcomes in K-12 mathematics prepared by the National Council of Teachers of Mathematics. In 1993, Project 2061's *Benchmarks for Science Literacy* appeared, and in 1996 the National Research Council of the National Academy of Sciences published *National Science Education Standards*, a document that includes recommendations for learning goals in science. Given the high level of consistency among the recommendations of these three groups—including tightened scope and increased emphasis on understanding, reasoning, and connectedness—a set of coherent goals for creating coherent curricula is readily available.

Developmental Sequence

For the ultimate science literacy envisioned in national goals to have any chance of being reached, K-12 benchmarks for learning along the way have to make developmental sense. One essential aspect of developmental coherence has to do with the logical dependence of complex ideas on precursor ideas. For example, the proposition that "force produces a change in motion" requires *some* prior understanding of what "force" and "change of motion" mean—but only some prior understanding, since part of understanding what "force" and "change of motion" mean involves knowing the relationship between them. Another aspect of coherence can be thought of as psychological, taking account of what preexisting notions students typically have with regard to given concepts and what difficulties they have in learning them. For instance, students tend not to distinguish among force, momentum, energy, pressure, power, and strength, and they tend to think of force as a property of objects rather than as a relation between objects.

Benchmarks for Science Literacy Chapter 15: The Research Base summarizes the cognitive research taken into account in assigning benchmarks to a particular grade range. In *Resources for Science Literacy: Professional Development,* this research summary is expanded further and electronically linked to the relevant sections of *Science for All Americans.*

Because a developmentally coherent set of learning goals must take such logical and psychological matters into account, cognitive scientists as well as teachers, scientists, mathematicians, and engineers were directly involved in the creation of *Benchmarks*. Three years of work beyond *Science for All Americans* on how student understanding and skills would grow with time went into *Benchmarks*. The main premise of the work was that what is learned now should be based on what was learned earlier, on what is capable of being learned now, and on what needs to be learned later. That rationale is readily seen in listings of benchmarks, as in the box on the facing page.

Thematic Connections

In principle, a coherent curriculum is one that makes conceptual sense at every level of instruction—from daily lesson plans to units, to courses, to the curriculum as a whole

A K-12 SET OF BENCHMARKS ON GRAVITY

By the end of 2nd grade, students should know that

• Things near the earth fall to the ground unless something holds them up.

By the end of 5th grade students should know that

• Things on or near the earth are pulled toward it by the earth's gravity.

• The earth's gravity pulls any object toward it without touching it.

By the end of 8th grade students should know that

• Everything on or anywhere near the earth is pulled toward the earth's center by gravitational force.

• Every object exerts gravitational force on every other object. The force depends on how much mass the objects have and on how far apart they are. The force is hard to detect unless at least one of the objects has a lot of mass.

• The sun's gravitational pull holds the earth and other planets in their orbits, just as the planets' gravitational pull keeps their moons in orbit around them.

By the end of 12th grade: students should know that

• On the basis of scientific evidence, the universe is estimated to be over ten billion years old. The current theory is that its entire contents expanded explosively from a hot, dense, chaotic mass. Stars condensed by gravity out of clouds of molecules of the lightest elements until nuclear fusion of the light elements into heavier ones began to occur. Fusion released great amounts of energy over billions of years. Eventually some stars exploded, producing clouds of heavy elements from which other stars and planets could later condense. The process of star formation and destruction continues.

• Life is adapted to conditions on the earth, including the force of gravity that enables the planet to retain an adequate atmosphere, and an intensity of radiation from the sun the allows water to cycle between liquid and vapor.

• Weather (in the short run) and climate (in the long run) involve the transfer of energy in and out of the atmosphere. Solar radiation heats the land masses, oceans, and air. Transfer of heat energy at the boundaries between the atmosphere, the land masses, and the oceans results in layers of different temperatures and densities in both the ocean and atmosphere. The action of gravitational force on regions of different densities causes them to rise or fall—and such circulation, influenced by the rotation of the earth, produces winds and ocean currents.

• Isaac Newton created a unified view of force and motion in which motion everywhere in the universe can be explained by the same few rules. His mathematical analysis of gravitational force and motion showed that planetary orbits had to be the very ellipse that Kepler had proposed two generations earlier.

• Newton's system was based on the concepts of mass, force, and acceleration, his three laws of motion relating them, and a physical law stating that the force of gravity between any two objects in the universe depends only upon their masses and the distance between them.

• General relativity theory pictures Newton's gravitational force as a distortion of space and time.

from *Benchmarks for Science Literacy* CHAPTER 4: THE PHYSICAL SETTING and CHAPTER 10: HISTORICAL PERSPECTIVES each year, to the curriculum over the years. Ideas and skills do not stand alone but are linked conceptually to other ideas and skills and appear in a variety of contexts.

This aspect of coherence, of course, is easier to achieve at the lesson-plan and unit levels than at higher ones. Indeed, the individual units making up a course of study are often very carefully organized, whereas the collection of units making up a course are largely independent of one another except for being in the same subject. Nevertheless, it is well within the reach of publishers to develop courses and even course sequences—blocks, in the language of earlier chapters of this book—in which the content is thoughtfully organized around a few pervasive themes, key ideas and skills are visited periodically (and with specific purpose) in different contexts, and their relationship is discussed explicitly. Crosscutting instructional themes can be used for creating such curriculum bridges. They can be concepts, applications, or historical episodes.

Conceptual themes—as the term is used in the COMMON THEMES chapters of *Science for All Americans* and *Benchmarks*—help us think about and understand ideas, processes, and events in many different subject-matter contexts. For example, the idea of "scale" has powerful applications in mathematics, physics, biology, engineering, agriculture, politics, etc., and so can be used to find bridge-building opportunities in the curriculum. An application theme is based on the notion that achieving understanding of an idea or acquiring a skill requires that students exercise the idea or skill in multiple contexts. Different subjects provide mutually advantageous opportunities for accomplishing this. For instance, science, social studies, and physical-education courses offer many occasions for students to apply newly learned mathematical ideas and skills, and, reciprocally, those subjects can easily provide mathematics classes with meaningful data to analyze and interesting patterns to model.

History is especially important in making cross-subject thematic connections. Science ought to have a major presence in history courses because of the enormous impact of science and technology on all of history. And history should be taken seriously in science courses because history alone provides a context for seeing how science really works over time and how it relates to mathematics and technology and to what else is happening in human culture.

So far in this chapter, we have briefly considered three separate aspects of curriculum coherence. In the following two sections, we use all three—literacy goals, developmental sequences, and thematic connections—to suggest ways to improve curriculum across grade levels and across subjects.

For more information on the use of themes, see the two subsections below entitled Within the Sciences and Among Science, Mathematics, and Technology.

IMPROVING COHERENCE ACROSS GRADE LEVELS

Surely one of the most common complaints about existing curricula is the repetition of topics that bore and discourage students. Yet one of the most effective means of strengthening a grasp of concepts and skills is to exercise them repeatedly in new contexts, at progressively higher levels of sophistication, and in relation to other concepts. At an extreme of informality, teachers can individually and spontaneously ask students to think again about something they have learned before in light of something that they have learned since. (Obviously even this informal approach requires teachers to be aware of the content of the curriculum in the prior years.) At a formal extreme, the curriculum can be carefully worked out, cooperatively by teachers at all the grade levels involved, to have built-in occasions for reflection on and consolidation of ideas.

Yet fostering coherence across grades is not without problems. Teachers often choose to use only some parts of instructional materials that were designed to cover a whole year or even several years. This practice puts pressure on materials developers to avoid instructional units that depend on students having experienced earlier parts of the program.

Sally Forth by Howard & Macintosh

Whatever the teacher's deliberate role in such selection may be, however, there is also an unavoidable discontinuity that results from students changing schools and school districts. One approach to reducing this disruption is to promote more uniform instructional programs—districtwide, statewide, or even nationwide—and more uniform grade placement of specific learning goals. The means of achieving these goals would be left substantially open. Another approach is to expand curricular and instructional alternatives, but keep track of students' achievements so that they can eventually fill all requirements.

Project 2061's experience has shown that cooperative planning across grade levels even as broad as primary to high school seems to have surprising benefits for teachers, in terms of their understanding their role in curriculum planning, and for students, in terms of how they experience the coherence of the curriculum. It is important to contrast cooperative K-12 planning to a more traditional approach in which (in effect) high-school teachers tell middle-school teachers who in turn tell elementary teachers what topics they have to cover. Much more productive is planning in which teachers work together to create a curriculum that includes revisiting certain topics at successively more sophisticated levels and in different contexts, but avoids needless repetition. (The revisiting, needless to say, should not be merely a reprise, but be focused on growth in understanding of specific aspects.) Suggestions for organizing such an effort and for placing and relating benchmarks follow in the next two subsections.

Using *Benchmarks for Science Literacy*

For a good start on increasing grade-level coherence in science, mathematics, and technology, teachers can use *Benchmarks for Science Literacy*. The first step is to identify all of the benchmarks to be addressed in the curriculum for some topic of modest size. When agreement has been reached on relevant benchmarks, decisions need to be made on which particular grades or subjects will or will not target them. Finally, the desired configuration of instruction is compared with the actual configuration in the current curriculum and plans are made to bring the two closer together.

Consider, for instance, the example on the following page of how this approach could play out using benchmarks for the topic of heredity. If it is difficult to assemble a committee of K-12 teachers, then teachers from different grades within the same grade ranges can form groups. Thus, a committee of middle-school teachers would see that there are three heredity benchmarks in grades 6-8, which means students are expected to learn the ideas somewhere between the beginning of the 6th grade and the end of the 8th grade. The group would decide which grade would be responsible for targeting each of those benchmarks, and agree that the other grades would desist—or deliberately reinforce them in a thoughtful way. In doing so, some attention would have to be paid to closely related benchmarks that appear under other headings. For example, the third benchmark for grades 6-8 in the *Benchmarks* section Heredity is closely related to one at the 6-8 level in the *Benchmarks* section Evolution of Life.

Committees of elementary and high-school teachers could undertake similar tasks by assigning the heredity benchmarks for grades K-2, 3-5, and 9-12. A better procedure would be for teams to be made up of teachers from adjacent grade ranges or, better still, from the entire K-12 span. In all cases, the task is to reach agreement on which grades are responsible for targeting the specified benchmarks.

ASSIGNING BENCHMARKS TO GRADES

A Middle-School Example for the Topic of Heredity

Benchmark	Assigned Grade
From *Benchmarks* section Heredity, p. 108:	
• In some kinds of organisms, all the genes come from a single parent, whereas in organisms that have sexes, typically half of the genes come from each parent.	Grade 6
• In sexual reproduction, a single specialized cell from a female merges with a specialized cell from a male. As the fertilized egg, carrying genetic information from each parent, multiplies to form the complete organism with about a trillion cells, the same genetic information is copied in each cell.	Grade 6
• New varieties of cultivated plants and domestic animals have resulted from selective breeding for particular traits.	Grade 7
From *Benchmarks* section Evolution of Life, p. 124:	
• Small differences between parents and offspring can accumulate (through selective breeding) in successive generations so that descendants are very different from their ancestors.	Grade 8

The committee might reason that teaching the first two benchmarks in grade 6 would allow placing human fertilization at the beginning of grade 7, where it could contribute to subsequent benchmarks in the Human Development section (and health education) on conception and embryo growth. The third benchmark would be needed as a precursor to the fourth, which might well be postponed to grade 8, when students might be more ready to consider its implications for evolution.

Teachers may be called upon to focus on benchmarks that heretofore they had not. In those instances, the team needs to agree on what is to be eliminated to make room (as described in Chapter 7). Before the beginning of the next cycle, the participating teachers should describe the experience and propose adjustments. Then another set of benchmarks should be selected and the process repeated.

Benchmarks provides further illumination of the coherence of ideas in brief essays that accompany the learning goals for each grade range. Consider the following excerpt from the Forces of Nature section:

Kindergarten through grade 2

The focus should be on motion and on encouraging children to be observant about when and how things seem to move or not move. They should notice that things fall to the ground if not held up. They should observe motion everywhere, making lists of different kinds of motion and what things move that way. Even in the primary years, children should use magnets to get things to move without touching them, and thereby learn that forces can act at a distance with no perceivable substance in between.

Grades 3 through 5

The main notion to convey here is that forces can act at a distance. Students should carry out investigations to become familiar with the pushes and pulls of magnets and static electricity. The term *gravity* may interfere with students' understanding because it often is used as an empty label for the common (and ancient) notion of "natural motion" toward the earth. The important point is that the earth *pulls* on objects.

Grades 6 through 8

The idea of gravity—up until now seen as something happening near the earth's surface—can be generalized to all matter everywhere in the universe. Some demonstration, in the laboratory or on film or videotape, of the gravitational force between objects may be essential to break through the intuitive notion that things just naturally fall. Students should make devices to observe the magnetic effects of current and the electric effects of moving magnets. At first, the devices can be simple electromagnets; later, more complex devices, such as motor kits, can be introduced.

Grades 9 through 12

Students should now learn how well the principle of universal gravitation explains the architecture of the universe and much that happens on the earth. The principle will become familiar from many different examples (star formation, tides, comet orbits, etc.) and from the study of the history leading to this unification of earth and sky. The "inversely proportional to the square" aspect is not a high priority for literacy. Much more important is escaping the common adult misconceptions that the earth's gravity does not extend beyond its atmosphere or that it is caused by the atmosphere....

Those comments can be extended conceptually by referring to other sections in *Benchmarks* that relate in interesting ways to the concept of gravity and its applica-

RESEARCH NOTES
page 340

Chapter **1** A THE SCIENTIFIC WORLD VIEW (universally by gravity)

8 B MATERIALS AND MANUFACTURING (properties of materials)

C ENERGY SOURCES AND USES (energy generation)

10 B UNITING THE HEAVENS AND EARTH

C RELATING MATTER & ENERGY AND TIME & SPACE (nature of gravity)

G SPLITTING THE ATOM (nuclear forces)

11 A SYSTEMS

C CONSTANCY AND CHANGE

D SCALE

◁ ALSO SEE

tions. The ones found in the "Also See" box at the beginning of the Forces of Nature section are shown above.

Inspecting the cognitive research findings found in *Benchmarks* CHAPTER 15: THE RESEARCH BASE will also help when considering curriculum coherence. Those for Forces of Nature are reproduced here:

The earth's gravity and gravitational forces in general form the bulk of research related to Forces of Nature. Elementary-school students typically do not understand gravity as a force. They see the phenomenon of a falling body as "natural" with no need for further explanation or they ascribe it to an internal effort of the object that is falling (Ogborn, 1985). If students do view weight as a force, they usually think it is the air that exerts this force (Ruggiero et al., 1985). Misconceptions about the causes of gravity persist after traditional high-school physics instruction (Brown & Clement, 1992) but can be overcome by specially designed instruction (Brown & Clement, 1992; Minstrell et al., 1992).

Students of all ages may hold misconceptions about the magnitude of the earth's gravitational force. Even after a physics course, many high-school students believe that gravity increases with height above the earth's surface (Gunstone & White, 1981) or are not sure whether the force of gravity would be greater on a lead ball than on a wooden ball of the same size (Brown & Clement, 1992). High-school students have also difficulty in conceptualizing gravitational forces as interactions. In particular, they have difficulty in understanding that the magnitudes of the gravitational forces that two objects of different mass exert on each other are equal. These difficulties persist even after specially designed instruction (Brown & Clement, 1992).

Using Strand Maps

See the draft strand map on page 184 of Chapter 6. Labels identify the kinds of information the maps provide. Many additional maps have been brought together in the Project 2061 *Atlas of Science Literacy*.

In *Benchmarks*, pointers to related ideas in other chapters are indicated by the "Also See" references. Strand maps display connections both over time and across topic areas. The maps depict the convergence of prior ideas into a new idea, the development of a conceptual strand through different stages of sophistication, and the linkage of several maps through ideas they share. Coherence is emphasized further by identifying "story lines" in strand maps that develop somewhat separately and eventually converge conceptually. Moreover, the maps are backed up with source materials from *Science for All Americans*, *Benchmarks* essays, and the research summaries in *Benchmarks* and *Resources for Science Literacy*. Maps also indicate connections to other maps. For example, the gravity map on page 184 shows connections to maps for climate and motion.

The prose in *Science for All Americans* helps to provide context for a map. For example, the coherent fabric of understanding that the gravity map depicts is articulated in *Science for All Americans* in the section Forces of Nature:

> Everything in the universe exerts gravitational forces on everything else, although the effects are readily noticeable only when at least one very large mass is involved (such as a star or planet). Gravity is the force behind the fall of rain, the power of rivers, the pulse of tides; it pulls the matter of planets and stars toward their centers to form spheres, holds planets in orbit, and gathers cosmic gas and dust together to form stars. Gravitational forces are thought of as involving a gravitational field that affects space around any mass. The strength of the field around an object is proportional to its mass and diminishes with distance from its center. For example, the earth's pull on an individual will depend on whether the person is, say, on the beach or far out in space.

It is important to understand that strand maps are not intended as literal prescriptions for organizing instruction, as if benchmarks could be neatly taught or achieved one at a time. Maps can help developers and designers attend to constraints on sequence and to keep track of what could be going on in students' minds as they experience successive lessons or units. The box that follows provides an example of how the configuration of instructional units can be related to key learning goals at different grade levels. The diagram on page 250 shows how those same instructional units also can be related to strand maps.

BUILDING CURRICULUM COHERENCE WITH RELATED INSTRUCTIONAL UNITS

An example of reflecting on coherence across grade levels is provided by a set of four actual instructional materials identified by Project 2061's curriculum-materials evaluation as having some potential to help students achieve some important but difficult benchmarks. All four units were developed in a research and development effort involving Michigan State University and the Michigan State Department of Education. Each unit requires about six to eight weeks of instruction (and so is on the small side for curriculum blocks).

The Lives of Plants, which could be taught as early as grade 5, focuses chiefly on the manufacture of sugar by plants (considering carbon dioxide, oxygen, water, and sugars as substances, not as molecules). *Matter and Molecules*, designed for as early as grade 6 but more likely to work better in grade 7, focuses on the particulate nature of matter and its differing arrangement and motion in solids, liquids, and gases. With that understanding, students are later able to study *Food, Energy, and Growth* in grade 8 or above, which follows the molecular fate of food in digestion and metabolism. Students' understanding of *Matter and Molecules* also enables them to take on *Chemistry That Applies* in grade 8 or above, in which conservation of mass is explicated as an unchanging number of atoms during chemical reactions, as they recombine into different molecules.

The diagram below shows how the units are conceptually related to one another and to other ideas (which would need to be treated in other units). For example, these units would depend somewhat on familiarity with living cells as a precursor to the extensive role that cells play in *Food, Energy, and Growth*. The grand payoff in terms of a molecular account of food webs would come still later, drawing on both *Food, Energy, and Growth* and *Chemistry That Applies*, on other aspects of *The Lives of Plants*, and on earlier familiarity with facts about what eats what.

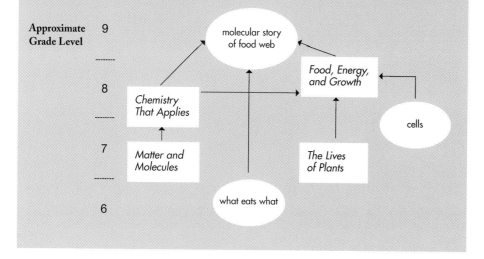

RELATING INSTRUCTIONAL UNITS TO STRAND MAPS

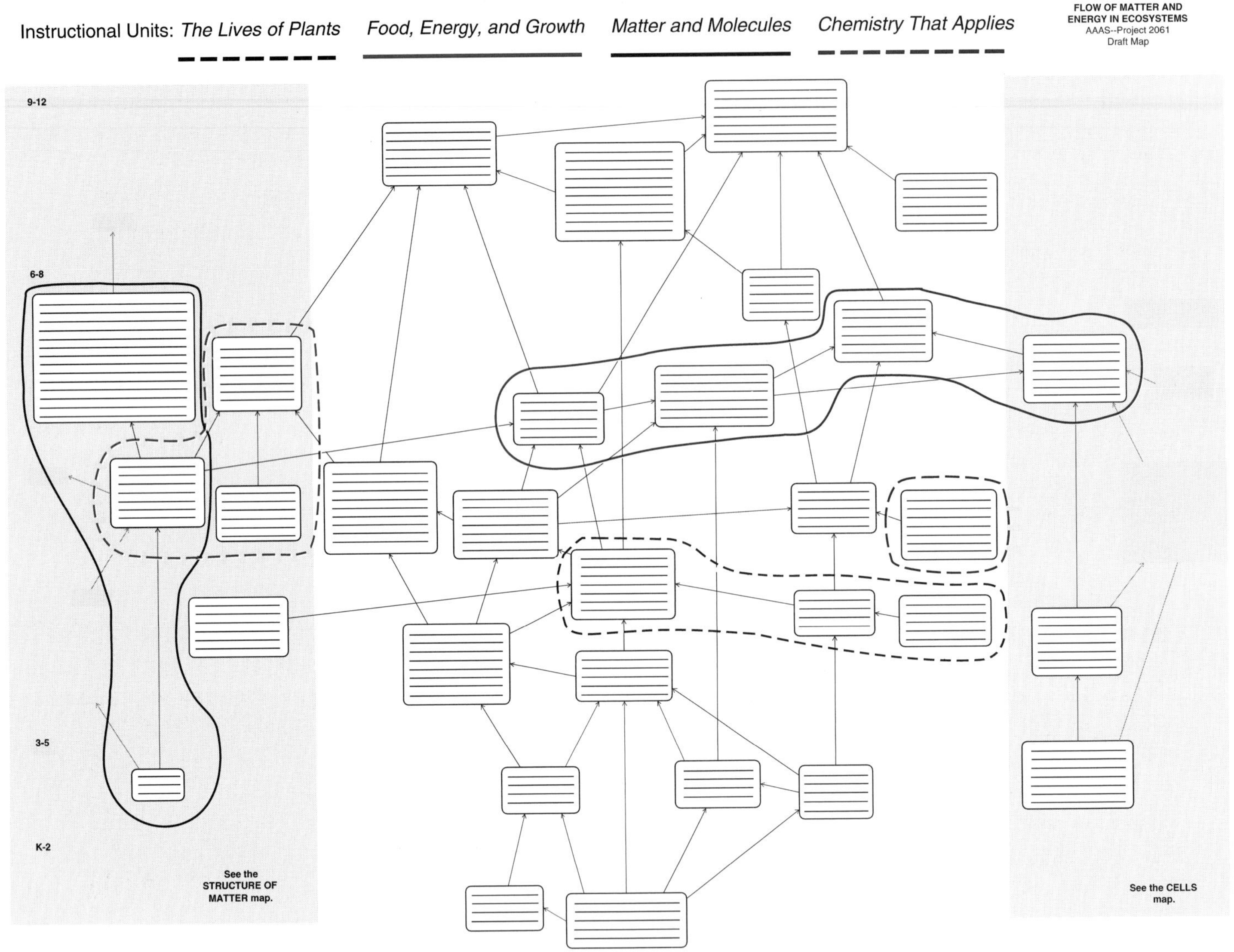

Instructional Units: *The Lives of Plants*   *Food, Energy, and Growth*   *Matter and Molecules*   *Chemistry That Applies*

FLOW OF MATTER AND ENERGY IN ECOSYSTEMS
AAAS--Project 2061
Draft Map

9-12

6-8

3-5

K-2

See the
**STRUCTURE OF
MATTER** map.

See the **CELLS**
map.

This diagram—derived from a composite of several maps from *Atlas of Science Literacy*—illustrates how instructional units can be related to strand maps. Loops on the maps enclose the benchmarks served by each unit. The *Matter and Molecules* unit serves a generally "vertical" set of learning goals, addressing a sequence of ideas over time. On the other hand, the *Lives of Plants* unit and the *Food, Energy, and Growth* unit serve a mostly "horizontal" set of goals, addressing a variety of ideas at approximately the same grade level. More than one unit can address the same idea—often a benefit if it is a particularly difficult idea.

In setting up this kind of intense study of strand maps, it may be difficult to put together a group to consider a map in its full K-12 range. A good first step is to create some pairs: K-2 teachers in a school can pair up with grade 3-5 teachers in the same school in addressing part of a growth-of-understanding sequence; or grade 3-5 teachers in an elementary school can pair with grade 6-8 teachers in a middle school; and similarly between a middle school and a high school. (The different groups need not all take on the same map.) The following year, each of these partial strands can expand up or down a grade range, and so on in each subsequent year. But if not right away, then as soon as possible, strand maps should be used by K-12 teams to ensure developmental coherence across the entire curriculum.

As teachers reach agreement on distributing the map's benchmarks among the grades and subjects, the assignments should be described and entered in a computer file that can readily be found and accessed by others. After the first implementation of the plan, a debriefing can help to identify problems of articulation that arose and to make improvements in the grade sequencing. That information, too, should be recorded and filed accessibly. Each year, additional strands can be implemented as teachers become comfortable with the process.

A reason to be circumspect about organizing lessons one-to-one on benchmarks is that often a benchmark contains several ideas that have similar content, but are not necessarily learned well at the same time. In strand maps, a single benchmark may be partitioned into separate boxes or even restated with a slightly different focus, to show more clearly its relationship to other ideas.

Designs on Disk provides suitable computer file forms in a database for assigning benchmarks to grade levels.

IMPROVING COHERENCE ACROSS SUBJECTS

There is no doubt that people learn, make sense of, and retain new concepts better when they are able to make clear connections to other ideas they are learning or already have, even if none of the ideas is completely clear. Sometimes that means instruction should relate ideas to students' own experience, or that instruction should draw students' attention to similar ideas in different subjects, or that students should exercise the ideas in a variety of contexts outside a particular course. Moreover, students' eventual ability to apply ideas usefully in new situations will benefit from their having previously practiced applying the ideas in a variety of new situations. Many educators seem to count on students eventually being able to make valuable cross-connections by themselves; but there is little evidence that students put ideas from different courses together. Actual integrated courses may not necessarily be an efficient way for students to learn helpful cross-connections; the research on student learning in integrated settings is still sparse. At the least, however, one-subject courses should provide opportunities for students to relate new ideas to other subjects.

Some divisions between subjects are based on their historical independence, some on beliefs about how students best learn, and some may seem to have been shaped,

It's plotted out. I just have to write it."

"If students are expected to apply ideas in novel situations, then they must practice applying them in novel situations. If they practice only predictable exercises, or unrealistic 'word problems,' then that is all they are likely to learn."
—*Science for All Americans, p.199*

almost accidentally, by school tradition. It is entirely possible to have a curriculum in which there is essentially no commerce between subjects—and in fact that is often the case. For over a century there have been calls to break down those curricular walls, often on the philosophical basis that the world itself has no such divisions, less often on the psychological basis of how students learn. In the future, it may be possible to control the extent of curriculum interconnectedness by selecting appropriate instructional blocks from a pool of well-described candidates that range from strictly discipline-based to fully integrated courses. Until then, headway toward increasing subject-matter coherence—within each science, among the sciences, and between them and other subjects—can be made in ways outlined below.

Within the Sciences

At any grade level, an examination of the science offerings will reveal opportunities to make some useful connections. Of the possibilities suggested here, the first two are interdisciplinary in nature, the next one has to do with shared attributes, and the last one is thematic.

The natural sciences. The separation of physics, chemistry, biology, and earth and space science at the secondary level is a powerful tradition. Many high-school science teachers take their professional identity from the specific discipline they teach (in contrast to high-school mathematics teachers, who are less inclined to make such distinctions). Science teachers in the middle grades are less attached to a particular science discipline, but even so, most middle-school general-science courses cycle through separate blocks of physical, biological, and earth sciences. Even courses described as general or integrated science may be episodically disciplinary—physics topics one day or week, biology topics another day or week, and so forth.

This is the order of study in the "Chicago" plan referred to in Chapter 2, p. 89.

For generations, there has been argument about how the natural-science disciplines should be ordered in the high-school curriculum. The most popular order is based on the mathematical sophistication required (as the courses are usually designed): Earth and space science, biology, chemistry, physics. It is argued from time to time (as each generation of teachers and scientists rediscovers the insight) that a more logical order would be on the basis of conceptual dependency: biology after chemistry (because biology draws on chemistry), and chemistry after physics (because chemistry draws on physics), giving the order physics-chemistry-biology, with earth science and astronomy introduced anywhere along the line, depending how explicitly they deal with physics principles.

But it may well be that neither of these sequences is altogether satisfactory. Surely it is not the case that all of physics should come before all of chemistry or all of chemistry before all of physics; there are *parts* of chemistry that could well come before parts of physics, and vice versa. Similarly, some parts of biology may well come before and some parts after chemistry and physics. Perhaps the ideal schedule would stretch all science subjects out over time, with ideas sequenced in a way that would allow them to be taught when they are needed and learnable, without regard to discipline, but with the disciplines supporting each other. But until such coordination can be engineered, coherence can be enhanced by pointing to some of the connections among the sciences, first in the elementary programs to lead students to *expect* connections, then within each high-school or middle-school science course. Strand maps are a good source of clues about when ideas outside a topic are relevant or necessary.

An achievable goal would be for each science course to include at least one unit that explicitly deals with connections to one of the other science disciplines. Yet developing such units would be challenging for teachers who have not experienced this sort of instruction themselves, either as students or as teachers. The simplest approach in high school, perhaps, is to import a teacher from another subject to teach one or a few lessons. A chemistry teacher, for example, could be invited into a biology course to introduce the nature of proteins and the action of enzymes. A biology teacher could reciprocate and expand on the role of enzymes for the chemistry students. An improvement on that would be for the biology and chemistry teachers to plan together what the specific learning goals for the lesson should be and how students' understanding can be assessed. In either case, the teachers will want to confer afterward about how it went and how it could be done better next time.

A more elaborate operation would be to have a committee of science teachers agree on some exchanges of content and help one another develop the units, assemble the materials, and the like. The teachers would keep a record of what happened and report back to the committee on their experiences working with the students. Over time, an interweaving of science subject matter could develop, even though the courses remain as separate disciplines.

The social sciences. The domain of curriculum labeled "social studies" often contains scientific elements such as psychology, anthropology, sociology, economics, and political science, as well as citizenship, civics, American and world history, and reflections on problems of society and political systems. Of the scores of learning goals stipulated in the National Council for Social Studies' (NCSS) *Curriculum Standards for the Social*

The National Science Teachers Association's *Scope, Sequence and Coordination* curricula have been a prominent attempt to coordinate high school science.

For efficiency, some strand maps are organized around topics that naturally draw on several disciplines—for example, "Flow of Matter and Energy in Ecosystems."

Designs on Disk contains a listing of *Science for All Americans* sections relevant to all ten NCSS strands. (A similar comparison also appears in *Resources for Science Literacy: Professional Development.*)

Studies, about 40 percent are addressed by both *Science for All Americans* and *Benchmarks for Science Literacy,* mostly in CHAPTER 7: HUMAN SOCIETY. For example, the first of the ten NCSS strands, "Culture," deals with the tendency for people from different cultures to arrive at different interpretations of information and experiences; benchmarks relevant to this idea occur in *Benchmarks* and *Science for All Americans* sections 1b: Scientific Inquiry, 6d: Learning, 7a: Cultural Effects on Behavior, and 12e: Critical Response Skills. Similar relationships can be found in the *National Science Education Standards (NSES)* sections Science and Technology, History and Philosophy of Science, and Science in Personal and Social Perspectives. From this area of common interest, science and social-studies teachers can work together to fashion units for each of their courses. The other social-studies strands provide similar opportunities.

Shared attributes. The individual science disciplines differ from one another in important ways, including their histories, interests, techniques, and languages. Yet they have much in common philosophically and operationally and borrow ideas, findings, and techniques from one another liberally. Some of the important shared attributes of the sciences are described in CHAPTER 1: THE NATURE OF SCIENCE and CHAPTER 12: HABITS OF MIND in both *Science for All Americans* and *Benchmarks.* Readings germane to those two chapters can be found on the *Resources for Science Literacy: Professional Development* CD-ROM.

Because science curricula rarely include study of the nature of science or general habits of mind, these areas are equally unfamiliar to all faculty and therefore invite collaboration. One way to proceed is for groups of teachers of science to begin with a single attribute—a subsection from the Scientific Inquiry section of CHAPTER 1: THE NATURE OF SCIENCE or from the Communication section in CHAPTER 12: HABITS OF MIND. For instance, the group might select this idea—"Science is a blend of logic and imagination"—for emphasis in all science courses. Teachers would agree to highlight it in the high-school physics unit on energy transformations, the high-school chemistry unit on periodic properties of elements, the high-school biology unit on heredity, the high-school earth-science unit on tectonic plates, the middle-school physical-science unit on force and motion, and an elementary-school science unit on diversity of life. In each of these contexts, there is a notable historical episode of a scientist—including such key individuals as Watt, Mendeleev, Mendel, Wegener, Newton, Darwin—coming up with a theory imaginatively, refining it logically, and substantiating it with empirical evidence.

As in other curriculum-related activities, discussion and study lead to recorded decisions, followed by development of instructional plans, implementation of the plans, debriefing, moving on to other topics and participation in another cycle.

Themes. Another way for the science curriculum to reduce the seeming isolation of the science subjects is for each course to draw on some common themes from time to time. "Theme" is often used in education as a synonym for "topic," as in "The fifth graders studied the theme of 'food' for three months, including farming; cooking; digestion; and stories, paintings, and songs about food." Organizing the curriculum this way may or may not be helpful, but for Project 2061, the terms "common themes" or "cross-cutting themes" are used to mean something very different. Themes are not artful headlines for collections of loosely related topics, rather they are underlying ways of thinking that cross disciplinary boundaries and prove fruitful in making sense of phenomena that appear to be quite diverse. In the *Science for All Americans* and *Benchmarks* CHAPTER 11: COMMON THEMES, thematic ideas are presented under four headings: Systems, Models, Constancy and Change, and Scale, each with several subheadings. (This organization is neither unique nor necessarily complete, and the subheadings could be arranged differently or even treated as distinct themes in their own right.)

Working in the collaborative manner outlined earlier, high-school teachers of earth and space science, biology, chemistry, and physics, middle-school general-science teachers, and science-interested elementary teachers as well, can select one theme to build into their respective courses. It is important not merely to provide examples of the theme in each course, but also to make explicit the connections among them. Another crucial step not to be overlooked is to select what content will be removed from each course to make room for emphasis on the theme. As the usual cycle of debriefing, modification, and expansion continues and confidence builds, the results should be shared with colleagues in the school district and even written up for publication in appropriate journals such as *The Science Teacher, American Biology Teacher, Journal of Chemical Education, Physics Teacher,* and *Journal of Geological Education.*

Among Science, Mathematics, and Technology

Science literacy calls for an understanding of the scientific enterprise as a whole, not just of the natural-science disciplines. In the Project 2061 view, this understanding includes science, mathematics, technology, and their interconnections. Hence a curriculum that claims to address science literacy needs to find a way to illuminate some

Most, but not all, of these historical episodes appear in *Science for All Americans* CHAPTER 10: HISTORICAL PERSPECTIVES.

In *NSES,* themes similar to these appear under "Unifying Concepts and Processes" at the front of each grade-range content recommendation.

For easy reference, CHAPTER 11: COMMON THEMES from *Science for All Americans* is reproduced on *Designs on Disk,* along with examples of where each of the *Science for All Americans* themes can be introduced into science courses.

of the vital connections between science and mathematics, science and technology, and technology and mathematics. This goal can be accomplished by exchanges of content, the use of common themes, and drawing on history.

Content exchange. Exchanges can be carried out by separate pairs of teachers—science/mathematics, mathematics/technology, technology/science—or by a group of teachers representing all three domains. The basic idea is to use *Science for All Americans* and *Benchmarks* to identify contexts in the curriculum in which subject matter can be shared naturally and with relative ease.

Ideally, the exchanges will go in both directions. For example, when middle-school mathematics teachers focus on different forms of graphing, they could use data provided by the science and technology teachers. Alternatively, the mathematics students could undertake science or technology activities that would generate real data for graphing. In turn, the science and technology teachers would agree to have those same students apply their developing graphing skills in laboratory and shop activities. (Even better, the teachers would plan together a study that would serve both subjects well.) Record keeping and debriefing would proceed as before as the number of exchanges grows gradually and without major disruptions of the curriculum.

The situation is often different in the elementary grades because of the absence of departments and the prevalence of self-contained classrooms. Many teachers in the early grades do not feel comfortable enough in science and mathematics to stray far from the textbooks, and may not delve into technology at all. Still, they have greater flexibility in the use of time and they are less hemmed in by discipline traditions. Since mathematics is regarded as a high-priority subject in elementary education, a productive approach is for a committee of interested teachers to look for science and technology activities that provide interesting occasions for students to apply and extend the mathematics they are learning. There is no shortage of such activities. The School Science and Mathematics Association has amassed a large collection of such activities over many decades. Several publishers have developed a variety of interdisciplinary resources, including the Lawrence Hall of Science's *GEMS* units and the AIMS Education Foundation's activities. The task in considering any of these materials is, of course, to select ones that lead toward specific learning goals for all the subjects.

Themes. The themes from *Science for All Americans* and *Benchmarks* that are useful for making connections within the natural sciences—models, systems, constancy and

Additional materials can be found in the *NSES* content standards section Science and Technology and in its Program Standard C on coordinating science and mathematics.

The School Science and Mathematics Association has published several collections of materials in its monograph series Classroom Activities and Topics for Teachers (see Bibliography).

change, and scale—are equally useful for linking science, mathematics, and technology. Although they are sophisticated concepts that may find their greatest value in the upper grades, modest beginnings toward them (as suggested in *Benchmarks*) can be made quite early.

After agreeing on a theme, a small group of teachers from different subject areas can identify contexts for developing, illustrating, or making explicit the theme. For the theme "models," biology students could examine the practice in medical research of using mice and other laboratory animals as stand-ins for humans, technology students could explore the ways in which scale models are used in architecture and manufacturing, and mathematics students could be introduced to the idea of mathematical models using computers. Students should reflect on all of these activities together, looking for specifics of how they are alike and different in the ways models are used.

History of science, mathematics, and technology. History is potentially a powerful unifier of science (natural as well as social), mathematics, and technology education. History provides many examples of how technological innovations have fostered whole new lines of scientific research, how scientific knowledge has led to the development of entirely new technologies, and of how mathematics has played an essential role in the advancement of scientific disciplines and technological fields.

Generalizations about these interdependencies will mean little to most students, however, unless they are generously supported by concrete, understandable examples. The episodes in *Science for All Americans* CHAPTER 10: HISTORICAL PERSPECTIVES provide some examples—such as linking germ theory to geometric growth of populations and to public sanitation—and should therefore be seriously considered when looking for ways to bring greater conceptual coherence to the high-school science curriculum.

With Other Subjects

In his famous 1959 lecture "The Two Cultures and the Scientific Revolution," C. P. Snow argued that there is a large and widening chasm between the sciences and humanities. Not all observers are as convinced as Snow was of the danger of the cultural separation, but few would claim that a reconciliation is near at hand. Still, a well-balanced curriculum committed to general education for all students can find ways to show that the various domains of learning are not remote islands of human thought and action, and indeed they can inform and complement each other. Eventually, it may happen that the pool of curriculum building blocks available to teachers will make it possible for curriculum designers to select some blocks in which imaginative developers have found ways to do that.

Science and general history. *Science for All Americans* recommends that student learning goals include knowledge of some specific episodes in the history of science. Two reasons are given: (1) without concrete examples, generalizations about how the scientific enterprise operates are empty slogans, however well they may be remembered; (2) some episodes in the history of the scientific endeavor are of surpassing significance to our cultural heritage, being milestones in the development of thought in western civilization.

The *Science for All Americans* recommendations, however, do not specify how those episodes in the history of science are to be taught, or by whom. A strong case can be made that high-school science and history teachers ought to share the responsibility. The problem with this prescription is that usually neither science teachers nor history teachers are well prepared in the history of science, few of them ever having taken even one course in it. Moreover, with few exceptions, science textbooks treat history superficially if at all, and history textbooks largely ignore the impact of science on the world.

Nevertheless, a collaboration between history and science teachers could well lead in time to some fruitful curriculum connections—on the one hand, to presenting science in a context of social history and, on the other hand, to getting the science correct and complete enough, and at an appropriate developmental level. A good way to begin would be for a group of science and history teachers in a high school—or elementary teachers who have comparably different specializations—to examine one of the ten *Science for All Americans* history-of-science episodes in detail. The purpose should be to first understand the material and then decide on how to teach it. References to and descriptions of background reading can be found in the Science

Peanuts by Charles M. Schulz

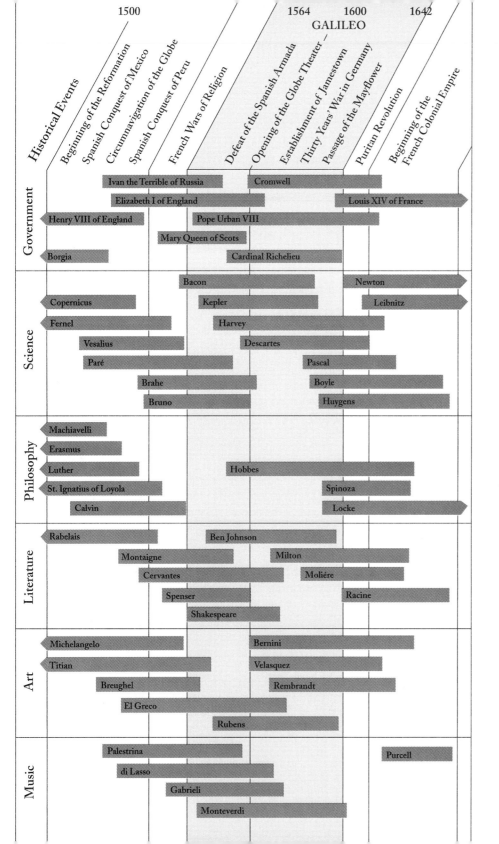

Chart of major cultural figures, in multiple fields, around the time of Galileo. This chart in the *Project Physics* high-school textbook was one of six, which centered on 300 B.C., and on 1500, 1600, 1700, 1800, and 1900 A.D.

Section headings for *Science for All Americans* CHAPTER 10: HISTORICAL PERSPECTIVES suggest the significant discoveries and changes that have shaped scientific knowledge:

- Displacing the Earth from the Center of the Universe
- Uniting the Heavens and Earth
- Relating Matter & Energy and Time & Space
- Extending Time
- Moving the Continents
- Understanding Fire
- Splitting the Atom
- Explaining the Diversity of Life
- Discovering Germs
- Harnessing Power

Trade Books component of the *Resources for Science Literacy: Professional Development* CD-ROM and also on *Designs on Disk.*

Once there is comfort with the episode on both sides, the study group should take up the instructional and curricular issues. Obviously the discussion should take account of the standards that have been written for history that would be relevant to that episode. After deciding which aspects of the episode will be introduced in which courses, the responsible teachers can then develop instructional plans to share with the group. Conceptually, different aspects of an episode can be treated in different courses, if scheduling allows the same group of students to be reached. The usual sequence of debriefing, modifications, and expansion should follow. The box on the previous page is an example of a time chart taken from a physics textbook that shows some possibilities for using historical events or figures to make cross-subject connections.

Science and social studies. Social-*studies* courses are usually not social-*science* courses, but there is every reason to call for greater interaction between the two. Social studies is becoming ever more quantitative, borrowing social-science methods and data (from sample surveys and demographic studies, for example) and using mathematics (especially statistics and probability, tables and graphs) to analyze data and display findings.

Drawing on the national mathematics education standards and on *Benchmarks* (CHAPTER 9: THE MATHEMATICAL WORLD and CHAPTER 12: HABITS OF MIND) and following the earlier described pattern of study, development, implementation, debriefing, and modification, a team of mathematics and social-studies teachers can begin introducing useful connections in both courses. For example, a mathematics teacher could use census and survey data from a social-studies course in teaching some aspect of data analysis—say, alternative ways to describe groups statistically. And a social-studies teacher could agree to have students use mathematical techniques, perhaps graphing, to look for relationships between variables from a mathematics course. Descriptions of outcomes, of course, would be made available to everyone in the school district.

In parallel, groups of social-studies teachers from as many different grades as possible can start with *Benchmarks*, one group using CHAPTER 7: HUMAN SOCIETY, another group CHAPTER 8: THE DESIGNED WORLD, and a third, the Critical Response Skills section from CHAPTER 12: HABITS OF MIND. In each case, the purpose would be to see if there are ways in which the ideas in *Benchmarks* can be used to flesh out some of the social-studies goals and strengthen the ties between the natural and social sciences. In return, science teachers could collaborate with social-studies teachers to devote some attention to social issues involving natural science, such as genetic engineering, costs of research, and environmental quality. This collaboration need not lead to team teaching but should involve cooperation in identifying topics and working out an instructional plan.

Science and literature. Science and English teachers can begin by exploring possibilities for collaboration without making a commitment in advance to introduce changes in their courses. They could read and discuss a book such as Swift's *Gulliver's Travels*, Wells's *The Time Machine*, Ibsen's *An Enemy of the People*, Camus' *The Plague*, Brecht's *Galileo*, or Vonnegut's *Galapagos* to see if they could reach agreement on how a science background could improve student responses to literature and vice versa. Science fiction (other than purely space melodrama) often raises interwoven issues of technology and society or uses a technological excuse to speculate about societal possibilities. Alternatively, the groups could go over benchmarks from the *Benchmarks* sections The Scientific Enterprise, Health Technology, or Values and Attitudes to see what literary works they bring to mind. After trying out some of their ideas, the group could prepare reading lists for students of various age levels recommending works of fiction that are relevant to both science and English, along with suggestions for teachers on how to use such readings.

Science and physical education. Physical education has long had a place in the curriculum for its presumed ability to foster lifelong habits of good health. In the lower grades, the connection between exercise and good health is easy to make because only one or two teachers are involved, but in the upper grades it is not uncommon for physical-education teachers and life-science teachers to collaborate informally. Unfortunately, physical education is rarely exploited in behalf of mathematics education—a lost opportunity.

Because it is a natural source of quantitative data, physical education provides an excellent and interesting context for helping students practice mathematics and become aware of its applications in everyday life. In a cooperative venture for which data collection is carefully planned, students could keep data on themselves (height, weight, temperature, heart rate, etc.) and their performance in different sports events (running different distances, jumping, throwing, lifting, etc.). As data accumulate, they can be analyzed and graphed to show distributions, correlations, trends, and cycles. And if students also keep records of how much they practice, or other factors that they believe affect their performance, the analysis becomes all the more interesting. Individual information can be transferred anonymously to a collection so that

CLOSE TO HOME   JOHN McPHERSON

In an effort to emphasize both physical fitness and academics, officials at Westbury High devised aerobic algebra.

population data can be used in mathematics classes to develop skills in data analysis, without possible embarrassment to individual students. Ideally, computers would be available in the gym so that students could easily record data in their own secure files for later (private or collective) use; but computers are not absolutely essential.

Science and the arts. Using *Benchmarks for Science Literacy* and *National Standards for Arts Education*, teams of science teachers and of teachers in each of the arts (dance, music, theater, and the visual arts) can collaborate to identify connections. Concrete connections occur in the science of sound (for example, how wave patterns differ for different instruments, acoustics of theaters), the science of light (pigments, shadows, aerial perspective), and the science of motion (forces involved in balance, starting, and stopping). In CHAPTER 11: COMMON THEMES and CHAPTER 12: HABITS OF MIND of *Benchmarks*, the teams can seek higher-level connections with the arts standards related to creating, evaluating, and responding, and to making interdisciplinary connections. For example, they can introduce students to how modern technology has made possible new media and subjects for artistic expression or to how the ways in which scientists make scientific choices are similar to or different from the ways in which performing and visual artists make artistic choices. A similar process can be used for identifying connections between the arts goals and the mathematics goals in *Benchmarks* (or the national mathematics standards). The teachers should collaborate, as needed, in planning and in teaching, as well as in evaluating the effectiveness of their instruction.

Science and work. One reason that science, mathematics, and technology education is important to the future of students is that the scientific enterprise is the source of so many kinds of jobs in modern societies today and will in all likelihood be more so in the future. Too many students believe that science in this broad sense is the exclusive province of people with advanced university degrees. As students become interested in what they want to be when they grow up, teachers have the responsibility to inform them of the vast array of different kinds of interesting work (most of which are not at the Ph.D. level) that is possible for them if they keep up in their study of science and mathematics. Teachers should not try to push all students into science-related careers but should do what they can to help students keep their options open as long as possible.

One factor in relating science literacy to work is that many leaders in business have come to believe that career education is much too narrowly focused on particular jobs in the present. More and more, such leaders assert, vocational education fails to meet the entry demands for jobs that have a future—and a rapidly changing future at that. "Training," it has been said, is for when you know exactly what people will be doing; "education" is for when you don't. In the ever more rapidly changing workplace, specific school training for highly specific jobs makes less sense. If more and more jobs require flexible and generalizable skills—such as the ability to reason, communicate, cooperate, learn new systems, and ask and find answers to questions—then the best vocational training starts to sound very much the same as the best general education.

Another factor in relating science literacy to work is that science and mathematics courses in the upper grades of high school focus so strenuously on what is seen as preparation for college; undecided students or those headed into the workforce after graduation avoid them. The curriculum has been such in most school districts that fewer than half of all high-school graduates have taken courses in chemistry, physics, and mathematics beyond algebra. Science and mathematics courses that are typically strongly academic in tone may not even prepare students well for college, and rarely point to the full range of jobs that science and mathematics make possible even without a college degree.

Teachers of vocational subjects can join with teachers of science and mathematics to improve the curriculum for all students. The former can introduce more science and mathematics in their courses, and they can encourage vocational education students to continue taking academic courses. Science and mathematics teachers can introduce career applications and information into their courses, and they can find ways to make those courses more interesting and accessible to students who may be headed to work after graduation rather than to college.

Some of the best experiments in bridging academic and career domains have been attempted by technology educators. The explicitly technological sections of both *Science for All Americans* and *Benchmarks*—CHAPTER 3: THE NATURE OF TECHNOLOGY and CHAPTER 8: THE DESIGNED WORLD—should provide opportunities for joint planning by science and technology teachers, following suggestions made earlier in this chapter. Also, CHAPTER 11: COMMON THEMES and CHAPTER 12: HABITS OF MIND include directly relevant material.

This chapter, which could well have been expanded into a book of examples, completes the current account of Project 2061 ideas about curriculum reform. In the Epilogue that follows, we revisit some of the potentially controversial highlights.

Claude Monet, *The Water-Lily Pond*, 1899.

EPILOGUE
ANOTHER LOOK AT *DESIGNS*

Readers may very well question some of the central premises of this book—or what they perceive them to be. For example, readers may have concluded that, in effect, *Designs* claims that:

- There is no curriculum crisis.
- A curriculum is like a Boeing 777 or a garden wall.
- Curriculum design consists of reducing much of current content.
- Teachers are not curriculum designers.
- Curriculum design is futile without the aid of computers.

Would Project 2061 support those claims? Yes—to some degree and with certain qualifications, as the following commentary tries to make clear. But remember, it is the intent of *Designs* to stimulate readers' discussion, not to insist that they agree.

A sustained curriculum revolution is needed.

Much of *Designs* is cast in time frames of a decade or more. Given the long-term perspective of Project 2061, this should not be surprising. International studies of student performance in science and mathematics show that curriculum improvements are badly needed right away, but changes that can be made quickly rarely have a significant or lasting impact. Ambitious curriculum reform simply cannot be had in a hurry, no matter how much we might wish it or how alluring the latest panacea appears to be. It is possible, however, to undertake long-term curriculum-reform measures that contribute to worthwhile improvements along the way. In 1993, when work on *Designs* began, this point was discussed in an article published in *2061 Today* and reproduced on the following page. Part III of *Designs* provides another example of how to deal with both short-term needs and long-term goals.

It may also be said that, in an important sense, *Designs* is less about what curricula should be like, or even about how badly reform is or is not needed, than it is about how to create the curriculum you want. Its fundamental notion is that design should focus intensely on outcomes and on local choices of means to achieve them. Once learning goals have been agreed upon and constraints made clear, then any curriculum

One way to pursue the conversation about curriculum is to access the Project 2061 Web site at www.project2061.org.

LONG TERM IN THE SHORT RUN

Here's our predicament: We want to improve education now, right now, for the sake of today's students and the future of the nation; yet we know that quick fixes invariably fail. Problems are immediate and concrete, good solutions are generational and speculative.

This dilemma is one that Project 2061 had to confront from the start, having declared itself to be "long-term." Under-the-gun teachers and administrators, school-board members, governors, and legislators are understandably less than thrilled with a reform proposition that *deliberately* plans to take 25 years or so to have full impact.

Not seeming to serve the immediate, urgent needs of practitioners is but one of the possible shortcomings of long-termness. Another is that it is difficult for long-term reform efforts to survive long enough to actually become long term. Given that resources for underwriting reform are in fact limited, it is difficult for foundations, government agencies, and industry to commit substantial funds year after year to projects whose payoff is over the horizon and, in the bargain, relatively uncertain. Moreover, it is vastly easier for a project to expound lofty goals and long-term strategies than to demonstrate that it is making acceptable progress in the here and now toward those goals. Faith has its limits, and adherents as well as doubters want evidence of progress sooner rather than later.

No matter, AAAS was not willing to back away from its determination to launch and sustain Project 2061, a truly long-term reform initiative in science, mathematics, and technology education. It made sense, however to take into account the traditional drawbacks of long-term

reform projects, ameliorating them at least, and turning them to advantage when possible.

The project's mid-range plans call for it, among other things, to have produced, field tested, revised, and disseminated a set of interrelated reform tools by the year 2000, fifteen years after its beginning, and to have provided training in their use for all educators who desire it. But in view of the desirability of contributing to immediate reform needs and of demonstrating near-term progress toward distant goals, Project 2061 decided to plan its work so that its products would emerge serially and, further, so that each would stand on its own as a useful resource for reformers.

Benchmarks for Science Literacy is a case in point. While *Benchmarks* was being developed, educators made extensive use of *Science for All Americans*, in the process reaching judgments on the quality and significance of the project's work. And now, as other Project 2061 products are being developed, educators can use *Benchmarks* along with *Science for All Americans* to further their current reform efforts and to make fresh judgments concerning the directions and value of the project and the feasibility of its long-term approach. In this way, each new product will provide immediate, practical help and also an opportunity to ascertain whether Project 2061 is still on course toward its ultimate goals.

The point is this: Long-term reform efforts can be designed to contribute significantly to near-at-hand improvements, but short-term efforts rarely contribute much to reform in the long run. Time counts—and so does timing.

that delivers the goods is acceptable—and whether it is deemed traditional or radical, discipline-based or integrated, takes a conceptual or an inquiry approach, is not the point. The teachers, consultants, and project staff who have had a hand in creating *Designs* have their own individual views, strongly held, on how urgently reform is needed and why, and on what curricula ought to be like ideally, but they have also understood that *Designs* is not the venue for expressing them.

A curriculum is like other design problems. In an important sense, curriculum is like the Interstate Highway System, a library, an assembly line, a chain of retail stores, a movie, a vegetable farm, a bank, the census, a research project, or a garden wall. All of these things are alike in that they are designed by people with specific goals in mind—they don't just happen. So it is with curriculum.

To declare that a curriculum is part of the designed world (in contrast to the natural world) scarcely seems radical. Still, some educators find that view to be too narrow and mechanistic. A jetliner, they might say, is a piece of machinery, a school is not; a movie is make-believe, education is real; an automobile assembly line turns out predetermined products, a curriculum helps individuals determine for themselves what they will become. Is there not something disturbing about applying architectural and engineering thinking to what is at heart an humanistic enterprise? Doesn't the quality of elementary and secondary education, like education at every level, depend more on the relationship between teachers and students than on which particular subjects are studied when?

There is little to be gained in trying to decide whether viewing curriculum design as one would airplane design is humanistic or mechanistic. Nor is there any reason why productive student-teacher relationships and a modern curriculum cannot coexist and reinforce each other. Indeed, the thrust of *Designs* is that it is possible to design curricula that are humanistic by any measure—the character of the experiences students have and the personal and social value of what they learn—but that such results are unlikely to happen accidentally. Quality instruction and quality curriculum are both essential and require planning.

In designing a brick wall, it can be assumed that all of the building blocks are essentially alike in substance and shape, whereas such an assumption cannot be made with regard to the building blocks of a curriculum. Perhaps designing a wall from irregular field stones would serve better, but in either case, all that is intended is to draw upon a familiar image to help think about curriculum design. A mason or

farmer creating a New England stone fence may reshape some of the stones, but mostly it is a matter of selecting and organizing stones; in the future as now, teachers will modify some of the blocks comprising their curriculum, but mostly it will be a matter of then selecting and organizing the blocks.

The other drawback is that, alas, the requisite pool of blocks does not now exist, and will take a decade or longer to create. That is, there are not blocks that meet the standards put forth in the Project 2061 curriculum-block template, and not a pool of blocks having the great variety called for in this book. But the idea here is to consider new possibilities for designing coherent K-12 curricula in the future—something very different from proposing a particular curriculum or curriculum approach. If the idea appears, after much debate, to have merit, perhaps steps will be initiated to make it possible; if not, others may be motivated to come forward with other possibilities.

Curriculum reform requires more than reducing the current content.
One chapter of *Designs* is devoted to the proposition that *to gain time to study key ideas thoroughly*, teachers simply should reduce the sheer number of different topics studied, trim unnecessary details from some topics, and cut back on the memorization of large numbers of technical terms. But if "simply" means easily, it is the wrong word, since few things in education are easily accomplished, and certainly not the measures suggested in CHAPTER 7: UNBURDENING THE CURRICULUM. If instead "simply" is taken to mean "only," it is equally misleading, for improvements in the curriculum cannot be achieved in isolation. Some of the issues that must be taken into account are discussed in *Blueprints for Reform* and summarized in the box on the facing page.

As is asserted at the beginning of this book, "unburdening" the curriculum ought not to be interpreted as "watering down." Perhaps the contrary is closer to what is intended: By removing that which is less important—certain topics, details, and terminology—time becomes available to learn better the ideas that are more important. The content specified in *Science for All Americans*, for example, was very carefully chosen to form a mutually supporting set of ideas that are also important and learnable.

Teachers should not be expected to design curriculum. Some readers may feel that *Designs* underplays the importance of teachers in curriculum design, others that it places unrealistic demands on teachers, and still others that the book is inconsistent with regard to the role of teachers in curriculum design. The project pleads innocent

BLUEPRINTS FOR REFORM: SCIENCE, MATHEMATICS, AND TECHNOLOGY EDUCATION

If lasting, meaningful reform of the science, mathematics, and technology curriculum is to occur, changes are needed throughout the entire K-12 education system. *Blueprints for Reform* examines 12 aspects of the system under three major themes:

THE FOUNDATION

Equity: How is the attainment of science literacy by all students impeded by policies and practices?

Policy: Do current local, state, and federal education policies help or hinder the realization of science literacy?

Finance: What are the costs, in terms of money and other resources, of "science literacy for all?"

Research: What kinds of research are needed to improve instruction for science literacy?

THE SCHOOL CONTEXT

School Organization: What will the realization of science literacy goals require of grade structure, teacher collaboration, and control of curriculum materials and assessments?

Curriculum Connections: How can connections among the natural and social sciences, mathematics, and technology be fostered?

Materials and Technology: What new resources are needed for teachers to help students become science literate?

Assessment: Do current assessment practices work for or against the kind of learning recommended in *Science for All Americans*?

THE SUPPORT STRUCTURE

Teacher Education: What changes are needed to produce teachers with the knowledge and skills to implement curricula based on science literacy goals?

Higher Education: What changes in admissions standards might be necessary to support K-12 reforms to promote science literacy?

Family and Community: How can families and communities help in supporting or implementing local, state, or national standards?

Business and Industry: In what ways can partnerships between business and education contribute to the attainment of science literacy?

BLUEPRINTS

FOR REFORM

SCIENCE, MATHEMATICS, AND TECHNOLOGY EDUCATION

AMERICAN ASSOCIATION FOR THE ADVANCEMENT OF SCIENCE

PROJECT 2061

"Too many states are assuming that, given a succinct vision statement or a curriculum framework, districts, schools, or teachers will be able to create the instructional materials they need.... This seems a poor assumption for any number of reasons, including the questionable premise that the vision is so well understood at the local level that it can readily be converted to specific curricula or materials. Also, quality control of new materials is required that is typically beyond the capability of individual schools."
—Zucker, et al., *Evaluation of the National Science Foundation's Statewide Systemic Initiatives Program*, 1995

on the first and third charges, but is ready to plea bargain on the second. Briefly summarized, *Designs* has this to say about the role of teachers in curriculum design:

Teachers are not, individually or collectively, in a position to develop the building blocks of the curriculum. They lack the time, technical support, training, and, for the most part, the desire to create whole courses and grade-level subject treatments, or the instructional materials to go with them. That is as it is now.

Teachers do have the responsibility—and would still have it in the plan put forth in this book—to modify curriculum blocks to fit local circumstances. To do that better, they will need better subject-matter and technical training than is currently the norm.

Teachers—but not all teachers—are absolutely essential in the development of curriculum blocks. They are needed to submit ideas for blocks, to work on block-development teams (whether on funded projects or with commercial publishers), to review draft material, and to carry out field studies. Teachers who participate in such work will become specialists but, except for periodic professional leaves to serve on development teams, will be mostly classroom based. It is classroom teachers, and not university scholars, writers, audiovisual and computer specialists, etc., who are most likely to keep curriculum-design efforts grounded in the reality of school life.

Once a proper pool of curriculum blocks exists, teachers will have a *central* role in curriculum design by assembly. Working in collaboration with parents, administrators, and others having a stake in the curriculum, groups of teachers will be responsible for formulating a design that meets the specifications set by the school board and is able to garner enough professional and public support to warrant implementation. These teachers will, of course, need to have had advanced training and be backed up with appropriate technical and clerical support.

Until such a pool of curriculum blocks exists, teachers can take the lead in improving the existing curriculum. They can begin to reduce curriculum overload and increase curriculum coherence (both developmentally and conceptually), and in the process incrementally improve the current curriculum and develop the skills and insights that will put them in the position to carry out enlightened curriculum design when the time comes. But for that to happen, teachers will need to receive authorization, encouragement, and professional support of a kind and level that is not common today.

Teachers can also attempt to affect the curriculum-materials market. They can demand, or at least favor, materials with substantive connections (not just the window-dressing of "topic" correspondence tables) to national and state standards.

Curriculum design requires the aid of computers. The concept of curriculum design by assembly and the role of computers in design are so closely related in *Designs* that the message of Part II could be taken by some readers to be that you can't do significant curriculum design without computers. As a practical matter, that is probably true, though in principle the operations could be carried out with paper and pencil. However, without electronic databases, rapid information exchange, analytic utilities, convenient record keeping, and the rest of the capabilities that computer-based information and communications systems bring to the task, sophisticated curriculum design would be enormously clumsy and time consuming—just as it would be in designing buildings, vehicles, gardens, molecules, and much else, were it not for computer-aided design.

Computers provide educators with an opportunity to do what heretofore has been pretty much out of the question—namely, to design highly sophisticated curricula that focus on specified learning goals and take local circumstances and preferences into account. The necessary computer and communications hardware exists right now, and Project 2061 will have developed the necessary software for search, selection, and record keeping within a few years (and is already testing some aspects of it).

Computers bring new possibilities to the task of curriculum design, but they do not bring solutions. If we are one day to have K-12 curricula that prepare all young people for interesting, responsible, and productive lives, it will only be because we are determined to have it so and are sufficiently creative and forward looking in our reform efforts. It will be up to teachers and those who prepare teachers, to parents, to citizens without children in the schools, to political and business leaders, to developers of instructional and assessment materials, to university researchers, and above all to education leaders. Can they come together on curriculum design? Will they? *Designs* is a modest attempt to move us all in that direction.

BIBLIOGRAPHY

References are listed under the chapters in which they appear. For convenience, the standards documents developed by various disciplines are grouped together at the end of the Bibliography.

SCIENCE LITERACY, CURRICULUM REFORM, AND THIS BOOK

Walker, D. (1990). *Fundamentals of curriculum.* San Diego, CA: Harcourt Brace Jovanovich.

PROLOGUE: DESIGN IN GENERAL

A survey of manufacturing technology. (1994). *The Economist, 330*(7853), 3-18.

Beam, W. R. (1990). *Systems engineering: Architecture and design.* New York: McGraw Hill.

Bucciarelli, L. (1994). *Designing engineers.* Boston: MIT Press.

Campbell, R. (1988, March). Learning from the Hancock. *Architecture, 77*(3).

Cowan, H. J. (1977). *The master builders: A history of structural and environmental design from ancient Egypt to the nineteenth century.* New York: John Wiley & Sons.

Cross, N. (1982). Designerly ways of knowing. *Design Studies, 3*(4), 221-227.

David, J. (1991). Restructuring and technology: Partners in change. *Phi Delta Kappan, 73*(1), 37-40.

French, M. J. (1988). *Invention and evolution: Design in nature and engineering.* Cambridge, England: Cambridge University Press.

Glegg, G. L. (1981). *The development of design.* Cambridge, England: Cambridge University Press.

Gregory, S. A. (Ed.). (1966). *The design method.* New York: Plenum.

Hopkins, H. J. (1970). *A span of bridges: An illustrated history.* New York: Praeger.

Hutchinson, J., & Karsnitz, J. (1994). *Design and problem solving in technology.* Albany, NY: Delmar Publishers.

Jones, J. C. (1992). *Design methods.* (2nd ed.). New York: Van Nostrand Reinhold.

Kuter, L. S. (1973). *The great gamble: The Boeing 747.* Tuscaloosa, Alabama: University of Alabama Press. A saga detailing Boeing and Pan Am's collaboration in producing the 747.

Lawson, B. (1990). *How designers think* (3rd ed.). London: The Architectural Press.

Leonhardt, F. (1984). *Bridges: Aesthetics and designs.* Cambridge, MA: MIT Press.

Maher, M. L., Balachandran, M. B., & Zhang, D. M. (1995). *Case-based reasoning in design.* Mahwah, NJ: Lawrence Erlbaum Associates.

McCullough, D. (1972). *The great bridge.* New York: Simon and Schuster.

McCullough, D. (1978). *The path between the seas: The creation of the Panama Canal, 1870-1914.* New York: Simon & Schuster.

Newhouse, J. (1982). A sporty game; III – Big, bigger, jumbo. *The New Yorker, 58*(19), 45-83. Describes attempts to design more profitable aircraft, culminating in the mid-sixties battle by airlines to preempt each other in strategic aircraft/engine purchases.

Petroski, H. (1994). *Design paradigms: Case histories of error and judgment in engineering.* New York: Cambridge University Press.

Petroski, H. (1982). *To engineer is human: The role of failure in successful design.* New York: St. Martin's Press.

Public Agenda. (1994). *First things first: What Americans expect from the public schools—Summary.* New York: Author.

Sobel, D. (1995). *Longitude: The true story of a lone genius who solved the greatest scientific problem of his time.* New York: Walker & Co.

Stites, J. (1994, Winter). W. Brian Arthur. *The Bulletin of the Santa Fe Institute, 9*(2), 5-8. The evolution of economics, according to Dr. Arthur.

Thompson, B. (1997). *Creative engineering design.* Okemos, MI: Okemos Press.

Chapter 1: CURRICULUM DESIGN

Anderson, J. (1994). Alternative approaches to organizing the school day and year. *The School Administrator, 51*(3), 8-11, 15.

Barzansky, B., & Gevitz, N. (Eds.). (1992). *Beyond Flexner: Medical education in the twentieth century.* Westport, CT: Greenwood Press.

Benathy, B. H. (1991). *Systems design of education: A journey to create the future.* Englewood Cliffs, NJ: Educational Technology Publications.

Breinin, C. (1994, Dec. 14). History learned is more important than history taught. [Letter to the editor]. *Education Week, 14*(15), 45.

Eisner, E. (1979). *The educational imagination.* New York: Macmillan.

Goodson, I. F. (1997). *The changing curriculum: Studies in social construction.* New York: Peter Lang.

Hutchinson, J., & Karsnitz, J. (1994). *Design and problem solving in technology.* Albany, NY: Delmar Publishers.

Kliebard, H. M. (1977). The Tyler rationale. In A. A. Bellack & H. M. Kliebard (Eds.), *Curriculum and evaluation* (pp. 57-67). Berkeley, CA: McCutchan Publishing. Kliebard explains and criticizes Tyler's 1950 classic about designing curriculum.

Kliebard, H. M. (1987). *The struggles for the American curriculum.* Boston: Routledge.

Kline, D. (1998, Spring/Summer). An overview of block scheduling. *Spectrum, 24*(1), 22.

Oxley, D. (1994, March). Organizing schools into small units: Alternatives to homogeneous grouping. *Phi Delta Kappan, 75*(7), 521-526.

Powell, A. G., Farrar, E., & Cohen, D. K. (1985). *The shopping mall high school: Winners and losers in the educational marketplace.* Boston: Houghton Mifflin.

Roth, K. (1994, Spring). Second thoughts about interdisciplinary studies. *American Educator, 18*(1).

Stephens, J. M. (1967). *The process of schooling: A psychological examination.* New York: Holt, Rinehart and Winston.

Thorndike, E. (1986). A neglected aspect of the American high school. *Educational Review, 33*(1907).

Tyler, R. W. (1950). *Basic principles of curriculum and instruction.* Chicago: University of Chicago Press.

Tyler, R. W. (1977). The organization of learning experiences. In A. A. Bellack & H. M. Kliebard (Eds.), *Curriculum and evaluation* (pp. 45-55). Berkeley, CA: McCutchan Publishing.

Tyson, H. (1997, July). *Overcoming structural barriers to good textbooks.* Paper presented at the meeting of the National Education Goals Panel, Las Vegas, NV.

Zajonc, A. (1992). Science within an ecology of mind: Alternatives in educational reform. *Holistic Education Review, 5*(3), 5-9. Several national studies find Americans' science literacy to be deficient. According to the author, three typical responses occur: national standards, high-tech educational tools, or "education as for-profit business."

Chapter 2: CURRICULUM SPECIFICATIONS

Anderson, J. (1995). *Who's in charge? State differences in public school teachers' perceptions of their control over determining curriculum, texts, and course content.* (Research Report AR 95-7007). Washington, DC: OERI, U.S. Department of Education.

Axtell, R., & Epstein, J. (1994, Winter). Agent-based modeling: Understanding our creations. *The Bulletin of the Santa Fe Institute, 9*(2), 28-29. This article states that key social structures, such as traffic patterns or even epidemics, emerge from exchange between separate agents.

Benjamin, H. (1939). *The saber-tooth curriculum by J. Abner Peddiwell, Ph.D.* New York: McGraw-Hill. Forward-thinking cavemen create curriculum based on real-world needs. Later, when those needs change, radicals insist curriculum adapt to meet current real-world needs.

Bracey, G. (1995). Variance happens—get over it! *Technos, 4*(3), 22-29. Bracey argues that some current education strategies, based on standards and outcomes that promote the expectation that all children can learn, are bound to fail.

Brett, M. (1996). Teaching extended class periods. *Social Education, 60*(2), 77-79.

Building new models for change in organizations. (1994, Winter). *The Bulletin of the Santa Fe Institute, 9*(2).

Burke, A. (1987). *Making a big school smaller: The school-within-a-school arrangement for middle level schools.* Unpublished manuscript.

Bybee, R. W., & McInerney, J. (Eds.). (1995). *Redesigning the science curriculum.* Colorado Springs, CO: BSCS.

Cahen, L. S., & Filby, N. N. (1979). The class size/achievement issue: New evidence and a research plan. *Phi Delta Kappan, 60*(7), 492-495, 538.

Checkley, K. (1995). Multiyear education: Reaping the benefits of "looping." *Education Update, 37*(8), 1, 3, 6. In "looping," a teacher will remain with a class for a minimum of two years, thus building stronger interpersonal relationships with the attendant benefits and some risks.

Glatthorn, A. A. (1994). *Developing a quality curriculum.* Alexandria, VA: Association for Supervision and Curriculum Development.

Goertz, M., Floden, R., & O'Day, J. (1996). *Systemic reform: Studies of education reform.* Washington, DC: OERI, U.S. Department of Education.

Huebner, D. (1975). The tasks of the curricular theorist. In W. Pinar. (Ed.), *Curriculum theorizing: The reconceptualists* (pp. 250-270). Berkeley, CA: McCutchan Publishing.

Labaree, D. F. (1999, May 19). The chronic failure of curriculum reform. *Education Week on the Web.* Available: www.edweek. org/ew/vol-18/36/abar.h18.

Managing change in education. (1998). Arlington, VA: Educational Research Service.

Macdonald, J. B. (1975). Curriculum theory. In W. Pinar. (Ed.), *Curriculum theorizing: The reconceptualists* (pp. 5-13). Berkeley, CA: McCutchan Publishing.

MacFarquhar, N. (1995, July 22). Trenton schools begin an experiment with year-round classes. *The New York Times,* pp. 21, 25.

Mitchell, C. T. (1993). *Refining designing: From form to experience.* New York: Van Nostrand Reinhold.

New schools from scratch. (1993). *High Strides, 5*(4), 7.

Ornstein, A. C. (1982, February). Curriculum contrasts: A historical overview. *Phi Delta Kappan, 63*(6), 404-408.

Perkins, D. N. (1987). Knowledge as design: Teaching thinking through content. In J. B. Baron & R. J. Sternberg, (Eds.), *Teaching thinking skills: Theory and practice* (pp. 63-85). New York: W. H. Freeman.

Pogrow, S. (1996). Reforming the wannabe reformers: Why education reforms almost always end up making things worse. *Phi Delta Kappan, 77*(10), 656-663.

Schmidt, W., McKnight, C., & Raizen, S. (1997). *A splintered vision: An investigation of U.S. science and mathematics education.* Norwell, MA: Kluwer Academic Publishers.

Schubert, W. (1993). Curriculum reform. In Cawelti, G. (Ed.), *Challenges and achievements of American education: The 1993 ASCD yearbook* (pp. 80-115). Alexandria, VA: Association for Supervision and Curriculum Development.

Schwab, J. J. (1983). The practical 4: Something for curriculum professors to do. *Curriculum Inquiry, 13*(3).

Sheingold, K. (1991). Restructuring for learning with technology: The potential for synergy. *Phi Delta Kappan, 73*(1), 17-27.

Stenvall, M. (1996). Year-round science: Shorter year-end breaks plus longer classes equals success. *The Science Teacher, 63*(6), 32-34.

Walker, D. F., & Schaffarzick, J. (1974, Winter). Comparing curricula. *Review of Educational Research, 44*(1).

Wilson, K., Daviss, B. (1994). *Redesigning education.* New York: Henry Holt.

Chapter 3: DESIGN BY ASSEMBLY

Barnes, R., Straton, J., & Ukena, M. (1996). A lesson in block scheduling. *The Science Teacher, 63*(6), 35.

Canaday, R. L., & Rettig, M .D. (Eds.). (1996). *Teaching in the block: Strategies for engaging active learners.* Larchmont, NY: Eye on Education.

Canaday, R. L., & Rettig, M. D. (1993). Unlocking the lockstep high school schedule. *Phi Delta Kappan, 75*(4), 310-314.

Christensen, C. R. (1987). *Teaching and the case method: Text, cases, and readings.* Boston: Harvard Business School.

Cooper, S. L. (1996). Blocking in success: Plan ahead for big dividends from a new schedule. *The Science Teacher, 63*(6), 28-31.

Day, M. M., Ivanov, C., & Binkley, S. (1996). Tackling block scheduling: How to make the most of longer classes. *The Science Teacher, 63*(6), 25-27.

Northwest Regional Educational Lab. (1990). *Literature search on the question: What are the advantages and disadvantages of various scheduling options for small secondary schools (high school and middle schools)?* Eugene, OR: Author.

Sizer, T. (1992). *Horace's school: Redesigning the American high school.* Boston: Houghton Mifflin.

Sommerfeld, M. (1996). More and more schools putting block scheduling to test of time. *Education Week, 15*(35), 1, 14, 15, 17.

Wiggins, G. (1989, November). The futility of trying to teach everything of importance. *Educational Leadership, 47*(3), 44-59.

Chapter 4: CURRICULUM BLOCKS

Adams, D. C., & Salvaterra, M .E. (1997). *Block scheduling: Pathways to success.* Lancaster, PA: Technomic Publishing.

Anderson, C. W., Roth, K. J., Hollon, R., & Blakeslee, T. (1987). *The power cell: Teacher's guide to respiration* (Occasional Paper No. 113). East Lansing, MI: Michigan State University, Institute for Research on Teaching.

Brearley, D., Ezell, D., Matthews, S., McGirr, B., Rossman, P., Sharp, R., Valenzuela, S., Vincent, F., & Welty, K. (1992). *Thoughts on design and writing design blocks.* Summer Institute conducted by Project 2061, Ithaca, NY.

Collins, A. (1991). The role of computer technology in restructuring schools. *Phi Delta Kappan, 73*(1). 28-36.

Doll, R. (1996). *Curriculum improvement.* (9th ed.). Needham Heights, MA: Allyn & Bacon.

Glegg, G. L. (1973). *The science of design.* Cambridge, England: Cambridge University Press.

Irmsher, K. (1996). Block scheduling in high schools. *Oregon School Study Council, 39*(6).

Jackson, P. W. (Ed.). (1992). *Handbook of research in curriculum: A project of the American Educational Research Association.* New York: Macmillan.

Jones, J. C. (1980). *Design methods: Seeds of human futures.* New York: Wiley.

Lindsay, J. (May 20, 1999). *The case against block scheduling.* Available: www.jefflindsay.com/Block.shtml

Luyten, H. (1994). *School size effects on achievement in secondary education: Evidence from the Netherlands, Sweden and the USA.* Paper presented at the annual meeting of the American Educational Research Association, New Orleans, LA.

Multiyear assignment of teachers to students. (1998). Arlington, VA: Educational Research Service.

Northeast and Islands' Regional Educational Laboratory at Brown University. (1997, September). *Block scheduling: Innovations with time.* Providence, RI: Author.

Orpwood, G., & Garden, R. (1998). *Assessing mathematics and science literacy.* Vancouver, Canada: Pacific Educational Press.

Queen, J. A., & Isenhour, K. G. (1998). *The 4x4 block schedule.* Larchmont, NY: Eye on Education.

Roberts, D., & Östman, L. (Eds.). (1996). *The many meanings of science curriculum.* New York: Teachers College Press.

Rossi, P. H., & Freeman, H. E. (1993). *Evaluation: A systematic approach.* Newbury Park, CA: Sage Publications.

Schroth, G. (1997). *Fundamentals of school scheduling.* Lancaster, PA: Technomic Publishing.

Selby, C. C. (1993). Technology: From myths to realities. *Phi Delta Kappan, 74*(9), 684-689.

Sykes, G. (1996). Reform of and as professional development. *Phi Delta Kappan 77*(7), 465-467.

Usiskin, Z. (1994, Winter). Individual differences in the teaching and learning of mathematics. *UCSMP Newsletter, 14,* 7-14.

Weiner, J. (1994) *The beak of the finch.* New York: Knopf.

Willis, S. (1993). Are longer classes better? *ASCD Update, 35*(3), 1-3.

Chapter 5: HOW IT COULD BE: THREE STORIES

American Association for the Advancement of Science. (1998). *Blueprints for reform.* New York: Oxford University Press.

Claxton, G. (1991). *Educating the inquiring mind: The challenge for school science.* Hertfordshire, England: Harvester Wheatsheaf.

Cuban, L. (1990). Reforming again and again and again. *Educational Researcher, 19*(1), 2-13.

Elmore, R. (1996, Spring). Getting to scale with good educational practice. *Harvard Educational Review, 1*(66), 1-26.

Fullan, M. (1993). *Change forces: Probing the depths of educational reform.* London: Falmer Press.

Fullan, M., & Pomfret, A. (1977, Winter). Research on curriculum and instruction implementation. *Review of Educational Research, 47*(1).

Gardner, H. (1992). The two rhetorics of school reform: Complex theories vs. the quick fix. *Chronicle of Higher Education, 38*(35).

Gee, W. D. (1997). The Copernican plan and year-round education. *Phi Delta Kappan, 78*(10), 793-796.

Gray, D. (1988, Summer). Socratic seminars: Basic education and reformation. *Basic Education: Issues, Answers, and Facts, 3*(14).

Hall, G. E. (1979, Summer). The concerns-based approach to facilitating change. *Educational Horizons.*

Herriott, R. E., & Gross, N. (Eds.). (1979). *The dynamics of planned educational change.* Berkeley, CA: McCutchan.

Hord, S., Stiegelbauer, S., & Hall, G. (1984, September-December). Principals don't do it alone: Researchers discover second change facilitator active in school improvement efforts. *R&DCTE Review,* II (3).

Lazerson, M. (1986). Review of "A Study of High Schools." *Harvard Educational Review, 56*(1).

Leithwood, K. A. (Ed.). (1982). *Studies in curriculum decision making.* Toronto, Ontario: The Ontario Institute for Studies in Education.

Miles, M. B. (1986, May). *Research findings on the stages of school improvement.* Unpublished manuscript.

Newmann, F., & Wehlage, G. (1995). *Successful school restructuring.* Madison, WI: Center on Organization and Restructuring of Schools.

Tyack, D., & Cuban, L. (1997). *Tinkering toward utopia.* Cambridge, MA: Harvard University Press.

Introduction to Part III: IMPROVING TODAY'S CURRICULUM

Petroski, H. & E. Kastenmeier. (1995). *Engineers of dreams.* New York: Knopf.

Chapter 6: BUILDING PROFESSIONAL CAPABILITY

Ability grouping: Effects and alternatives. (1998). Arlington, VA: Educational Research Service.

American Association for the Advancement of Science. (1997). *Resources for science literacy: Professional development.* New York: Oxford University Press.

American Association for the Advancement of Science. (in press). *Atlas of science literacy.* Author.

American Association for the Advancement of Science. (in press). *Resources for science literacy: Curriculum materials evaluation.* New York: Oxford University Press.

Anderson, C. (1991). Policy implications of research on science teaching and teachers' knowledge. In M. M. Kennedy (Ed.), *Teaching academic subjects to diverse learners* (pp. 5-30). New York: Teachers College Press.

Anderson, R. H., & Nelson, B. (1993). *Nongradedness: Helping it to happen.* Lancaster, PA: Technomic Publishing.

Arzi, H. J. (1988). From short- to long-term: Studying science education longitudinally. *Studies in Science Education, 15,* 17-53.

Bybee, R. (1993). *Reforming science education: Social perspectives and personal reflections.* New York: Teachers College Press.

Fullan, M. (1985). Change processes and strategies at the local level. *The Elementary School Journal, 85*(3).

Fullan, M., Galluzzo, G., Morris, P., & Watson, N. (1998). *The rise and stall of teacher education reform.* Washington, DC: American Association of Colleges for Teacher Education.

Gabel, D. L. (1994). *Handbook of research on science teaching and learning.* New York: Macmillan.

Goodlad, J. (1984). *A place called school: Prospects for the future.* New York: McGraw-Hill.

Grouws, D. T. (1992). *Handbook of research on mathematics teaching and learning.* New York: Macmillan.

Jackson, P. W. (Ed). (1992). *Handbook of research on curriculum: A project of the American Educational Research Association.* New York: Macmillan.

Jensen, R. J. (1993). *Research ideas for the classroom: Early childhood mathematics.* New York: Macmillan.

Johnson, B. (1996). *The performance assessment handbook: Portfolios and Socratic seminars.* (Vol. 1). Larchmont, NY: Eye on Education.

Millar, R., & Driver, R. (1987). Beyond processes. *Studies in Science Education, 14,* 33-62. Given that education is cyclical in its phases, the British authors examine the current and past emphasis on the methods of science as justification for its inclusion in the curriculum.

National Education Commission on Time and Learning. (1994). *Prisoners of time.* Washington, DC: U.S. Government Printing Office. This serves as a comprehensive review of how time is used in the public school system and its effect on student learning.

National Center for Education Statistics. (1993). *Time in the classroom.* (Research report NCES 94-398). Washington, DC: OERI, U.S. Department of Education.

National Staff Development Council. (1995). *Standards for staff development: Elementary school edition.* Oxford, OH: Author.

National Staff Development Council. (1995). *Standards for staff development: High school edition.* Oxford, OH: Author.

Newman, D. (1992). Technology as support for school structure and school restructuring. *Phi Delta Kappan, 74*(4), 308-315.

Nosich, G. (1991). The goals of science education. *Inquiry, 9*(1), 1, 4-6.

Owens, D. T. (1993). *Research ideas for the classroom: Middle grades mathematics.* New York: Macmillan.

Pate-Bain, H., Achilles, C. M., Boyd-Zaharias, J., & McKenna, B. (1992). Class size does make a difference. *Phi Delta Kappan, 74*(3), 253-256.

Patton, M. Q. (1997). *Utilization-focused evaluation.* (3rd ed.). Thousand Oaks, CA: Sage Publications.

Pumphrey, S. (1991). History of science in the national science curriculum: A critical review of resources and their aims. *The British Journal for the History of Science, 24.*

Roth, K. (1987). *Learning to be comfortable in the neighborhood of science: An analysis of three approaches to elementary science teaching.* Unpublished manuscript.

Shanker, A. (1995, February 5). *Where we stand: Disciplinary learning.* Available: www.aft.org/stand/previous/1995/020595.html

Slavin, R. E., Madden, N. A., Dolan, L. J., Wasik, B. A., Ross, S. M., & Smith, L. J. (1994, April). "Whenever and wherever we choose"—the replication of "Success for All." *Phi Delta Kappan, 75*(8), 639-647.

Sneider, C. (1972, September). A different "discovery" approach. *The Physics Teacher, 10*(6), 327-329.

Stavy, R. (1991, October). Children's ideas about matter. *School Science and Mathematics, 91*(6), 240-244.

Wilson, P. S. (1993). *Research ideas for the classroom: High school mathematics.* New York: Macmillan.

Chapter 7: UNBURDENING THE CURRICULUM

Bracey, G. W. (1992, June 12). Cut out algebra! *The Washington Post,* p. C5.

Carr, J. F., & Harris, D. E. (1993). *Getting it together: A process workbook for K-12 curriculum development, implementation, and assessment.* Needham Heights, MA: Allyn & Bacon.

Groves, F. H. (1995). Science vocabulary load of selected secondary science textbooks. *School Science and Mathematics, 95*(5), 231-235.

Schmidt, W., McKnight, C., & Raizen, S. (1997). *A splintered vision: An investigation of U.S. science and mathematics education.* Norwell, MA: Kluwer Academic Publishers.

Schools Brief: Tests of the truth. (1992). *The Economist, 325*(7785), 106-107. This features a brief look at the character and history of the scientific experiment as it is affected by its human proponents.

Chapter 8: INCREASING CURRICULUM COHERENCE

Berlin, D. (1991). *A bibliography of integrated science and mathematics teaching and learning literature.* Bowling Green, OH: School Science and Mathematics Association.

Berlin, D. (Ed.). (1997). *School science and mathematics integrated lessons.* Bloomsburg, PA: School Science and Mathematics Association.

Bybee, R. W., Powell, J. C., Ellis, J. D., Giese, J. R., Parisi, L., & Singleton, L. (1991). Integrating the history and nature of science and technology in science and social studies curriculum. *Science Education, 75*(1), 143-155.

Davison, D. M., Miller, K. W., & Metheny, D. L. (1995). What does the integration of science and mathematics really mean? *School Science and Mathematics, 95*(5), 226-230.

Dede, C. (1989). The evolution of information technology: Implications for curriculum. *Educational Leadership, 7*(1), 23-26.

Edling, W. (1992). *Creating a tech prep curriculum.* Waco, TX: Center for Occupational Research and Development.

Ellis, A.. K., & Stuen, C. J. (1998). *The interdisciplinary curriculum.* Larchmont, NY: Eye on Education.

Evans, D. (1995). Education for the workplace: Another form of elitism. *Education Week, 15*(10), 35.

Fogarty, R. (Ed.). (1993). *Integrating the curricula: A collection.* Palatine, IL: IRI/Skylight Publishing.

National Council on Economic Education. (1997). *Voluntary national content standards.* New York: Author.

Scarborough, J. D. (1993). Integrated models for teachers. *The Technology Teacher, 52*(5), 26-30.

Schmidt, W., Raizen, S., McKnight, C., Britton, E., Nicol, C., & Robitaille, D. (Eds.). (1993). *Curriculum frameworks for mathematics and science.* Vancouver, Canada: Pacific Educational Press.

Snider, R. (1992). The machine in the classroom. *Phi Delta Kappan, 74*(4), 316-323. This article chronicles the chaotic use of technology in the classroom and argues that humans direct the machinery, not the other way around.

Underhill, R. G. (1995). Integrating math and science: We need dialogue! *School Science and Mathematics, 95*(5), 225.

Walker, D., & Soltis, J. (1997). *Curriculum and aims.* (3rd ed.) New York: Teachers College Press.

Willis, S. (1995). Making integrated curriculum a reality. *Education Update, 37*(4).

Work-based learning models for school-to-work programs. (1995). *Education Daily, 28*(198), 6.

Epilogue: ANOTHER LOOK AT *DESIGNS*

Zucker, A. A., Shields, P. M., Adelman, N., & Powell, J. (1995). *Evaluation of the National Science Foundation's Statewide Systemic Initiatives (SSI) program: Second-year report, cross-cutting themes* (NSF Publication No. 96-48). Washington, DC: National Science Foundation.

Current national standards documents

American Association for the Advancement of Science. (1989). *Science for all Americans.* New York: Oxford University Press.

American Association for the Advancement of Science. (1993). *Benchmarks for science literacy.* New York: Oxford University Press.

Center for Civic Education. (1994). *National standards for civics and government.* Calabasas, CA: Author.

Consortium of National Arts Education Associations. (1994). *National standards for arts education: What every young American should know and be able to do in the arts.* Reston, VA: Author.

Council for Basic Education. (1998). *Standards for excellence in education: A guide for parents, teachers, and principals for evaluating and implementing standards for education.* Washington, DC: Association for Supervision and Curriculum Development.

Geography Education Standards Project. (1994). *Geography for life: National geography standards.* Washington, DC: National Geographic Research and Exploration.

International Technology Education Association. (in press). *Standards for technology: Content for the study of technology.* Reston, VA: Author.

Joint Committee on National Health Education Standards. (1995). *National health education standards: Achieving health literacy.* Reston, VA: Association for the Advancement of Health Education.

Mohnsen, B. (Ed.). (1998). *Concepts of physical education: What every student should know.* Reston, VA: National Association for Sport and Physical Education/AAHPERD.

National Center for History in the Schools. (1994). *National standards for history for grades K-4: Expanding children's world in time and space.* Los Angeles, CA: Author.

National Center for History in the Schools. (1994). *National standards for United States history: Exploring the American experience.* Los Angeles, CA: Author.

National Center for History in the Schools. (1994). *National standards for world history: Exploring paths to the present.* Los Angeles, CA: Author.

National Council for the Social Studies. (1994). *Expectations of excellence: Curriculum standards for social studies.* Washington, DC: Author.

National Council for Teachers of Mathematics. (1998). *Principles and standards for school mathematics—discussion draft.* Reston, VA: Author.

National Council of Teachers of English and the International Reading Association. (1996). *Standards for the English language arts.* Urbana, IL: National Council of Teachers of English.

National Research Council. (1996). *National science education standards.* Washington, DC: National Academy Press.

National Standards in Foreign Language Education Project. (1996). *Standards for foreign language learning: Preparing for the 21st century.* Lawrence, KS: Allen Press, Inc.

ACKNOWLEDGEMENTS

PROJECT 2061 STAFF

Designs for Science Literacy has been a collaborative effort of the entire Project 2061 staff, under the leadership of the project's founding director F. James Rutherford and its associate director Andrew Ahlgren. Former staff members who contributed to the *Designs* effort include Sofia Kesidou, Senior Research Associate; Gerald Kulm, Program Director; Alan Stonebraker, Technology Specialist; and Luli Stern, Research Associate.

George Nelson *Director*
Andrew Ahlgren *Associate Director*
Ryan Arndt *Project Assistant*
Vikas Arya *Technology Specialist*
Lucia Buie *Administrative Support Specialist*
Fernando Cajas *Research Associate*
Ann Caldwell *Research Associate*
Barbara Goldstein *Administrative Coordinator*
Laura Grier *Project Coordinator*
Terry Handy *Writer*
Cheryl Jackson *Administrative Support Specialist*
Mary Koppal *Communications Director*
Lori Kurth *Research Associate*
Lester Matlock *Project Administrator*
Francis Molina *Technology Director*
Kathleen Morris *Senior Program Associate*
David Peery *Technology Specialist*
Thelxiopi Proimaki *Technology Specialist*
Jo Ellen Roseman *Curriculum Director*
Susan Shuttleworth *Senior Editor*
Brian Sweeney *Webmaster*
Soren Wheeler *Project Coordinator*
Linda Williams *Senior Financial Analyst*

PROJECT 2061 PROFESSIONAL DEVELOPMENT PROGRAMS

Scott May *Executive Director*
Ayda Argueta *Administrative Support Specialist*
Mary Ann Brearton *Professional Development Manager*
Linda Hackett *Math Workshop Leader*
John Howe *Marketing Manager*
Regina Oglesby *Project Assistant*
Joseph Watson *Marketing Associate*

PROJECT 2061 SCHOOL-DISTRICT CENTERS

GEORGIA
Sue Matthews *Center Director*
PHILADELPHIA, PA
Natalie Hiller *Center Director*
SAN ANTONIO, TX
Joan Drennan-Taylor *Center Director*
SAN DIEGO, CA
Danine Ezell *Center Director*
SAN FRANCISCO, CA
Bernard Farges *Center Director*
WISCONSIN
Leroy Lee *Center Director*

REVIEWERS

Erma Anderson *National Science Teachers Association*
Frank Betts *Association for Supervision and Development*
Chris Castillo-Comer *Texas Education Agency*
Theron Blakeslee *Michigan Department of Education*
Gerald Bracey *Alliance for Curriculum Reform*
Joe Exline *Council of State Science Supervisors*

REVIEWERS *continued*

Eileen Ferrance *Northeast and Islands Regional Educational Laboratory at Brown University*

Deborah Fortune *American Association for Health Education*

Thomas Hatfield *National Art Education Association*

Henry Heikkinen *University of Northern Colorado*

Pat Heller *University of Minnesota*

Marlene Hilkowitz *Penn-Delco School District, PA*

Paul Kimmelman *Superintendent, West Northfield School District No. 31, IL*

Martharose Laffey *National Council for the Social Studies*

Kathleen O'Sullivan *San Francisco State University*

Harold Pratt *National Research Council*

Steven Rakow *University of Houston, Clear Lake*

Colleen Reeve *San Antonio Public Schools*

Rita Rice *School District of Philadelphia (retired)*

Doug Roberts *University of Calgary*

John L. Roeder *Teachers Clearinghouse for Science and Society*

Gary Scott *Resource Teacher, Los Angeles Unified School District*

James Shymansky *University of Iowa*

Cary Sneider *Boston Museum of Science*

Ginny Van Horn *Education and Human Resources, American Association for the Advancement of Science*

James Varnadore *San Diego, CA*

Decker Walker *Stanford University*

Sylvia Ware *American Chemical Society*

Robert E. Yager *University of Iowa*

Judith Young *National Association of Sport and Physical Education*

John Zola *The New Vista High School, Boulder, CO*

EDITORIAL

Barbara DeGennaro *Indexer*

Paul Elliott *Wordsworth Communications*

DESIGN AND PRODUCTION

Isely and/or Clark Design

Carol Hardy Design

Type Shoppe II Productions Ltd.

ART RESEARCH

Steve Diamond, Inc.

Photo Assist

INDEX

CREDITS

Installing *Designs on Disk*

In *Designs for Science Literacy* there are many suggestions for how school teams can approach the improvement of curriculum and instruction. The companion CD-ROM *Designs on Disk* provides a variety of aids for curriculum design—information resources, databases, and utilities—to help educators undertake these suggested activities. The CD-ROM also includes a fully searchable electronic version of the book's text and, to set a wider context or explain rationales, it provides links from the resources, databases, and utilities to the relevant text sections. In addition to the following brief directions for installing the disk, more detailed instructions and demonstrations on how to use the CD-ROM are provided on the disk itself.

SYSTEM REQUIREMENTS

To install *Designs on Disk*, you must have the following hardware and operating system:

**IBM or compatible computer system running Microsoft®
Windows® 95, 98, NT® 4.0, or 2000.**

4X or faster CD-ROM drive

Sound capabilities

Minimum memory of 32 Mb

Minimum available hard disk space of 60 Mb

Minimum screen resolution of 800 x 600 (optimum: 1024 x 768)

Color monitor set to 16-bit display (65,356 colors)

OR

Macintosh® Power PC (operating system 7.6 or later)

4X or faster CD-ROM drive

Sound capabilities

Minimum memory of 24 Mb

Minimum available hard disk space of 60 Mb

Minimum screen resolution of 832 x 624 (optimum: 1024 x 768)

Color monitor with Color Depth set to Thousands of Colors

Startup disk named "Macintosh HD"

For both platforms, the available hard disk space requirement will be lower if you already have some or all of the required software.

INSTALLATION

The *Designs on Disk* installer checks to see if your computer has all of the needed components and installs them if necessary. It also creates (for a Windows operating system) a shortcut on your desktop and an item in your Start > Programs menu or (for a Macintosh operating system) an alias on your desktop and an item in your Apple menu.

MICROSOFT WINDOWS

NOTE: *(If you are using a Windows NT or 2000 operating system, check first with your system administrator to make sure you have the proper privileges to install the software.)*

1. Before you begin, save your work and exit all open programs. If you are using Netscape AOL Instant Messenger, right click on its icon located next to the task bar clock, choose Exit, and then click Yes.

2. Insert the CD-ROM into your CD-ROM drive. An autorun program launches the *Designs on Disk* installer automatically. If autorun is disabled on your computer and the installer does not start automatically, use Windows Explorer to view the contents of your CD-ROM drive and double-click on Start.exe.

Windows 2000 users only: If the setup program detects that you are using Windows 2000 and that the preferred browser is not yet installed, you will get the following message:*"You are using Windows 2000 and the preferred browser is not yet installed. Please install the browser by looking at the contents of the* Designs on Disk *CD-ROM and double-clicking* *cc32e473.exe under the Windows\Netscape folder. Clicking the OK button below will exit this setup program so you can install the browser. Installation will resume after restart."*

Click OK to exit. View the contents of *Designs on Disk* using Windows Explorer. Double-click on the file cc32e473.exe under the Windows\Netscape folder. After the Netscape browser is installed, restart your computer. The setup program should resume. If it does not, go back to Step 2.

3. On the subsequent screens, accept the Software License Agreement, choose the default Typical Setup, and follow the prompts.

4. At the end of the installation, you will be given the options of viewing a demonstration movie and displaying a postage–paid registration form that you can fill in and send to Project 2061. Click Yes to accept these options. You may disregard the registration dialog if you are going to register online (see the Registration section at the end of these instructions).

5. After restarting your computer, you may have to create a Netscape user profile if the setup program installed this browser. Follow the prompts and if you are not sure about any of the information requested, please contact your system administrator or Internet Service Provider.

6. Double-click the shortcut on your desktop to start *Designs on Disk*.

MACINTOSH

1. Before you begin, save your work and exit all open programs.

2. Insert the CD-ROM into your CD-ROM drive. Double-click the *Designs* icon on your desktop.

3. Double-click on Mac_install and click Continue on the opening screen.

4. On the subsequent screens, accept the Software License Agreement, accept the default Easy Install, and follow the prompts.

5. Click Quit whenever you see the following dialog:

Installation was successful. If you are finished, click Quit to leave the Installer. If you wish to perform additional installations, click Continue.

Continue | Quit

6. If the setup program determines that the Shockwave and Flash plug-ins need to be installed, it will inform you which browser these components will be installed in. Make sure that the following dialog shows Netscape Communicator. Otherwise, choose Browse, find the

Netscape Communicator file on your computer's hard drive, and select it.

7. Click Quit if you see the following dialog:

8. At the end of the installation, you will be given the options of viewing a demonstration movie and displaying a postage–paid registration form that you can fill in and send to Project 2061. Click Yes to accept either option. You may disregard the registration dialog if you are going to register online (see the Registration section at the end of these instructions).

9. Restart your computer. You may have to create a Netscape user profile if the setup program installed this browser. Follow the prompts and if you are not sure about any of the information requested, please contact your system administrator or Internet Service Provider.

10. To start *Designs on Disk*, double click the alias on your desktop.

REGISTRATION

After installing the CD-ROM, you can register your copy online by clicking on the blue Project 2061 logo in the upper left corner of the *Designs on Disk* home page (requires Internet connection). Registration entitles you to free technical support by phone for a period of ninety (90) days from the date of purchase and to unlimited support via e-mail.